The Phone Book

Second Edition

A Money-Saving Guide to Installing or Replacing Your Own Telephone Equipment

By Gerald Luecke
James B. Allen

REVISED 2ND EDITION, 1997

PROMPT© Publications is a division of Howard W. Sams & Company, A Bell Atlantic Company, 2647 Waterfront Parkway, E. Dr., Indianapolis, IN 46214-2041.

This book was originally developed and published as *Installing Your Own Telephones* by Master Publishing, Inc., 7101 N. Ridgeway Avenue, Lincolnwood, IL 60645-2621.

International Standard Book Number: 0-7906-1133-3

Editor: Charles Battle, Natalie F. Harris
Cover Design: Phil Velikan
Photography: Christina Smith
Model: Shey Query

Acknowledgments:
All photographs not credited are either courtesy of the authors, Master Publishing, Inc., or Howard W. Sams & Company. All product illustrations, product names and logos are trademarks of their respective manufacturers. All terms in this book that are known or suspected to be trademarks or services have been appropriately capitalized. PROMPT© Publications, Master Publishing, Howard W. Sams & Company, and Bell Atlantic cannot attest to the accuracy of this information. Use of an illustration, term or logo in this book should not be regarded as affecting the validity of any trademark or service mark.

PRINTED IN THE UNITED STATES OF AMERICA

9 8 7 6 5 4 3 2 1

TABLE OF CONTENTS

		Page
	Preface	v
Chapter 1.	The Telephone and Its Operation	**1**-1, 20
Chapter 2.	Assessing What You Have	**2**-1, 12
Chapter 3.	Replacing Modular Telephones	**3**-1, 14
Chapter 4.	Replacing Old Style Telephones	**4**-1, 18
Chapter 5.	Running Interconnecting Cables	**5**-1, 16
Chapter 6.	Adding Telephones to a Modular System	**6**-1, 12
Chapter 7.	Adding Telephones to an Old Style System	**7**-1, 10
Chapter 8.	Business Installations	**8**-1, 8
Chapter 9.	Prewiring Installations	**9**-1, 18
Chapter 10.	Accessory Equipment Installations	**10**-1, 18
Chapter 11.	Installation Checks and Troubleshooting	**11**-1, 14
	Glossary	**G**-1
	Index	**I**-1

To Alexander Graham Bell (1847-1922)
who started it all.

PREFACE

This book is an installation guide for telephones and telephone accessories. It has been written with the goal of making it easy for the inexperienced person to install telephones, whether existing ones are being replaced or moved or new ones added. Since new telephone equipment will have the new modular connections, installations emphasize the modular type; however, adding and replacing telephones of old style vintage and how the old style systems are converted to modular also are a major part of this book.

The book begins by explaining the telephone system and its operation and by showing how to assess a presently installed telephone system so that a plan can be laid out before installations begins. Clear step-by-step instructions, fully illustrated, follow for replacing and adding telephones, first modular then old style. Mixed between these installation instructions are directions for installing telephone cable. Business installations, prewiring installations, accessory equipment installations and checking and troubleshooting conclude this book.

With this book, a minimum of tools available around the house, and readily available and reliable parts, you should be able to handle any telephone installation for home, apartment or small business. We wish you success — the rest is up to you.

G.L.
J.A.

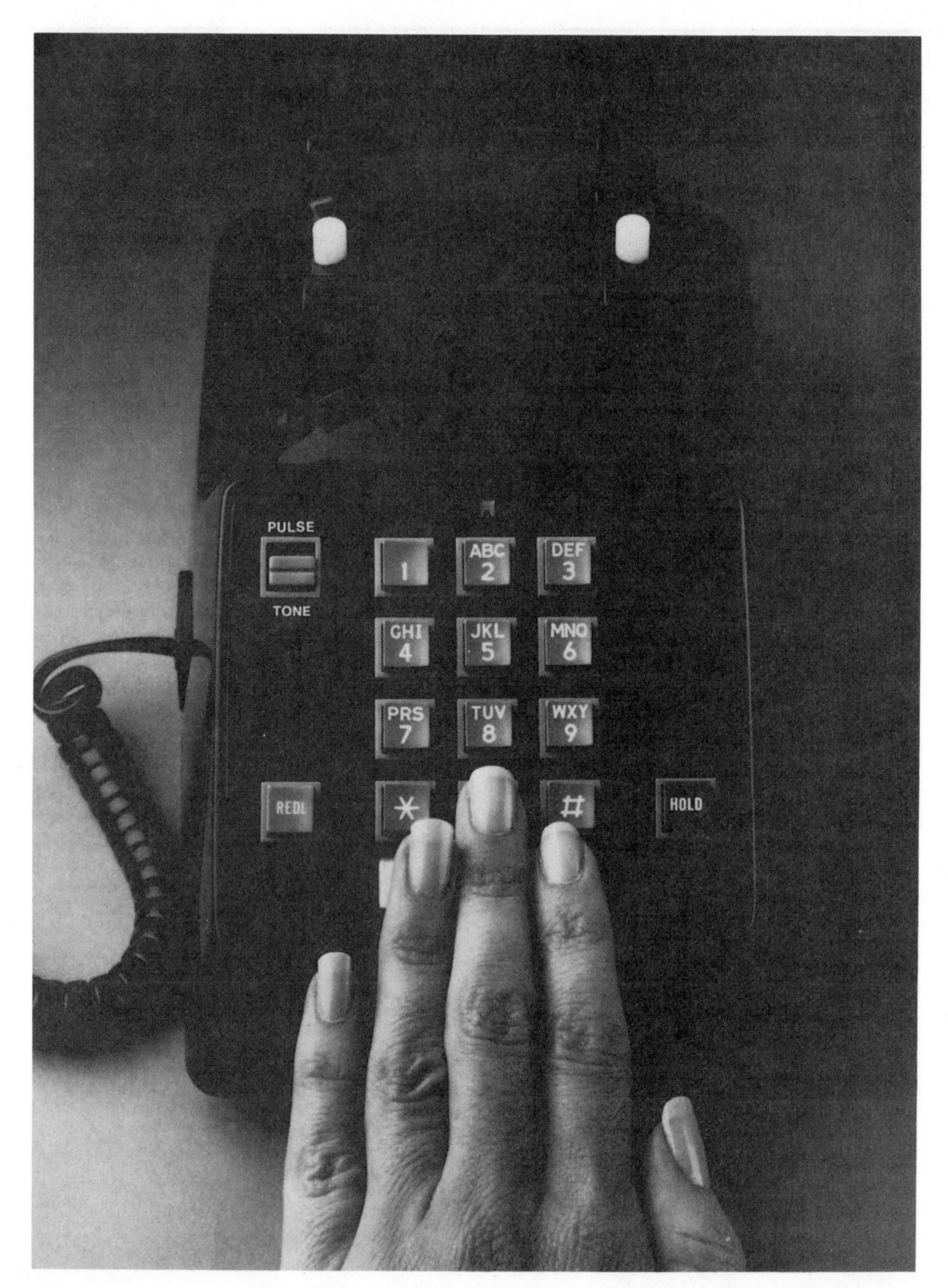
PULSE
TONE
1
ABC 2
DEF 3
GHI 4
JKL 5
MNO 6
PRS 7
TUV 8
WXY 9
REDL
*
#
HOLD

1-1

1 THE TELEPHONE AND ITS OPERATION

INTRODUCTION

Years ago it was not possible to install or even use equipment that was not the property of the local telephone company. As a result of a Federal Communications Commission ruling in 1977[1], those conditions have changed. Now it is not only possible, but also it is very practical, economical, and profitable (it saves money) for persons who live in an apartment; own or rent their own home; or run a small business to buy and install their own telephone equipment. In addition, it is not difficult to do. Especially with clear understandable instructions and reliable equipment.

That's the purpose of this book; to make it easy to install telephone equipment purchased from any of the conveniently available outlets. Based on highly visual instructions that proceed step by step, anyone, even though they are not mechanically or electrically inclined, can follow the book materials and successfully install wall-mounted, desk-mounted, portable extension telephones, and even accessory equipment. No extensive numbers of tools are required beyond that normally used in home maintenance.

REASONS FOR INSTALLING YOUR OWN TELEPHONES

There are a number of reasons for installing your own telephones. The primary reason is to save money, but in addition, doing your own installation gives you the flexibility of adding feature telephones and accessories that the telephone company may not offer. Your own installation allows you access to a telephone connection at any practical location that you desire. It also provides you with the opportunity to do your own maintenance.

WHAT ARE THE SAVINGS?

Table 1-1 lists examples of typical charges for equipment rentals, installations and service calls by the telephone company. A range of costs are shown to cover the variations that occur in different sections of the country.

[1]FCC Rules and Regulations, Part 68, *Connection of Terminal Equipment to the Telephone Network*, July 1977.

Table 1-1.
Telephone Company Charges

Condition	Type	Monthly	Installation
Home or Apartment Location (Single Line)	Rotary Dial Wall or Desk	$3-5	$50-90
	Push-Button Tone-Dialing Wall or Desk	$5-8	$50-90
	Specialty Telephone	$6-15	$50-90
Small Business Location (Single Line)	Rotary Dial Wall or Desk	$5-8	$70-100
	Push-Button Tone-Dialing Wall or Desk	$7-10	$70-100
	Specialty Telephone	$10-20	$70-100
Condition	**Type**	**Service Charge**	
Maintenance	Problem in Telephone Co. Equipment or Line	No Charge	
	Problem with Purchased Telephone	$50 and up	

Telephone companies that lease telephones charge a monthly rental fee for any of their telephones that are installed. This does not include any line or access charges which may be on your monthly bill. As shown in *Table 1-1*, the fee is $3 to $5 for a rotary dial wall or desk mounted telephone, $5 to $8 for tone-dialing push-button telephones and up to $15 for specialty telephones. Just by installing two telephones that replace existing telephone company units, $2 to $20 can be saved each month on rental fees. That's $24 to $240 per year! Even with an average cost of $50 for a telephone, your investment is paid back in two years.

SAVING EVEN MORE

More money can be saved when individuals do their own installation. *Table 1-1* shows that the telephone company would charge around $50 to $100 for installing a jack for an additional telephone. Still more money can be saved if you are capable of doing your own maintenance work; because residential service charges of $60 and up, and business service calls of around $80 and up, can be avoided.

Add to this the fact that if you move, you collect your telephones and take them with you. At the next location you need only pay for a hook-up charge and avoid costly installation charges.

WHAT ARE SOME OF THE AVAILABLE FEATURES

Specialty, Memory, Amplifiers

When you do your own installation you have the flexibility to choose among a wide variety of telephones and a wide variety of features. For example, instead of the standard rotary or push-button telephone, you may want a specialty phone like the one shown in *Figure 1-1*. Or a telephone like the one shown in *Figure 1-2* may suit your fancy. It has the capability to hold in memory frequently called telephone numbers with as many as 28 digits (enough to call overseas). Each number is available at the push of one button. Pulse or tone dialing is selectable, and there is the capability to dial without lifting the handset (on-hook dialing). In order to carry on completely "hands-free" conversations, this telephone is equipped with a speaker amplifier for two-way speech. It also has a liquid crystal display and a clock to time calls.

Cordless Telephones

If the need is for complete freedom of movement of the handset, then a cordless telephone like the one shown in *Figure 1-3* may be desirable. Or a small business with single line service may want to install an amplifier like that shown in *Figure 1-4* for conference calls. This same small business may need a unit like that shown in *Figure 1-5* to answer callers with a prerecorded message. All of these features are easy to install and can be added with very little effort.

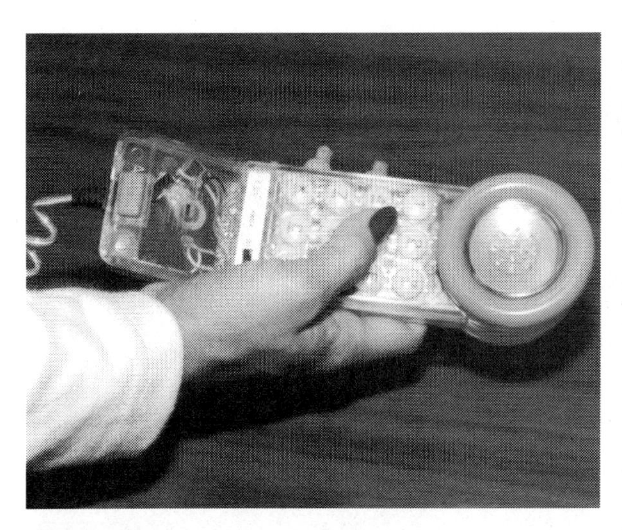

Figure 1-1.
Specialty Telephone

Figure 1-2.
Feature Telephone

WHERE DO YOU START?

Telephone companies use particular terms and definitions for the telephone sets, the functions they perform and the equipment (facilities) used to connect the calling party to the called party. So that you will better understand the installation instructions, let's review a number of these. With this understanding you will see how easy it is to do the installations. Let's look at the basic system first.

Figure 1-3.
Cordless Telephone
Courtesy of Southwestern Bell Freedom Phone®

Figure 1-4.
Conference Amplifier
Courtesy of Radio Shack

Figure 1-5.
Answerer with Digital Voice Announcement
Courtesy of Southwestern Bell Freedom Pnone®

BASIC SYSTEM

The telephone in your home, apartment or small business is connected, as shown in *Figure 1-6,* by a telephone line to a local telephone exchange called a *central office*. The telephone connection from your building to the central office is called the *local loop*. The telephone line actually is two wires called a *cable pair*. The pair of wires runs from the central office to your telephone one of the two ways shown in *Figure 1-6*. Either it is suspended on telephone poles and is attached to your house, apartment, or business building from overhead *(Figure 1-6a)* or it runs underground as shown in *Figure 1-6b*. The power for your telephone comes from the central office (indicated by the battery in *Figure 1-6a & b*).

Overhead Distribution

Many years ago, especially in the rural areas, individual telephone wires were strung on telephone pole cross bars so that the countryside looked like strings of a huge guitar. However, today, as shown in *Figure 1-6a*, many pairs of wires are bundled together in cables and suspended from poles. The poles may serve for electrical power distribution as well as telephone cable support. The overhead pair of wires that brings telephone service to your home, apartment, or business building connects at the nearest telephone pole to a pair of wires in the cable which brings the two-wire connection all the way from the central office.

Underground Distribution

The connection shown in *Figure 1-6b* is the same exactly except that the pair of wires runs underground from your property to a neighborhood *cable termination box* and then through underground cable to the central office. At selected locations additional cable termination boxes are placed along the route. The termination boxes are available either in manholes or on the surface. You certainly have seen telephone men working at these termination boxes up on poles or down in manholes. They are making telephone line connections; either installing new lines, modifying existing connections, or performing maintenance.

Outside Connection At Home, Apartment or Business

Whether the telephone line comes to your location overhead or underground, the pair of wires terminate at a *protector*. The protector does just what its name implies. It protects the pair of telephone wires into your building from high voltage spikes that might occur on the wires due to lightning bolts.

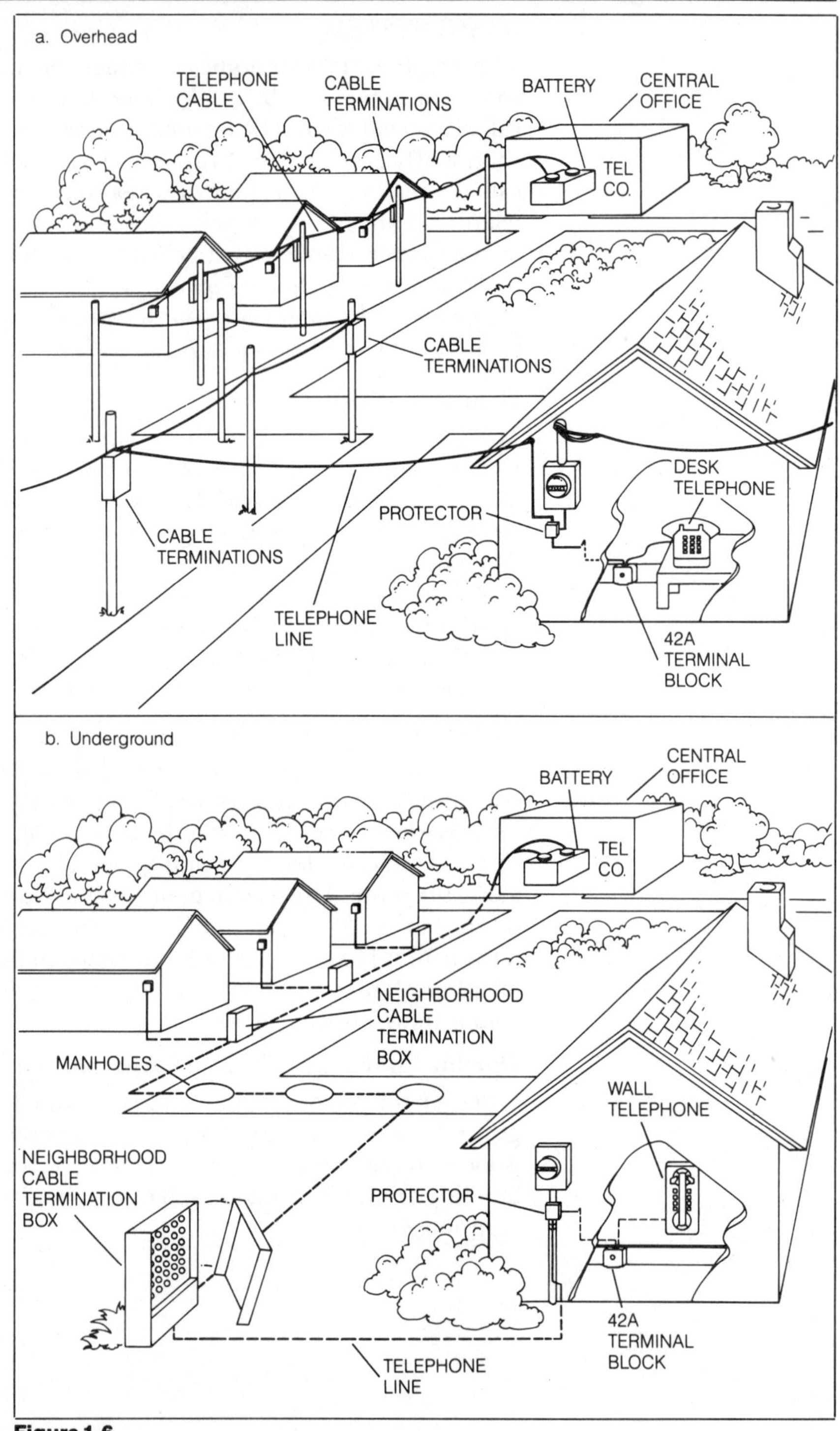

Figure 1-6.
The Telephone Line to the Central Office

A protector is shown in *Figure 1-7a*. It is used as a junction box. The nut and bolt terminals connect the cable pair from the central office to the cable pair that feeds telephone service into your home or small business. For apartments, townhouses and condominiums, the cable has many more pairs of wire than the single pair for a home or small business. Therefore, as shown in *Figure 1-7b*, a terminal strip with many protectors mounted together in an assembly is used to terminate the central office lines at an apartment, townhouse or condominium.

Inside Connection at Home, Apartment or Small Business

From the single protector, the telephone line feeds inside the house or small business. The line may run directly through the outside wall as shown in *Figure 1-6*, or it may feed through the roof overhang (soffit) into the attic, or it may feed through the outside wall into the basement. For apartments, wire pairs are patched from the protector terminals to multiple terminal strips that connect to wire pairs feeding each apartment.

a. Protector with Wires Connected to it

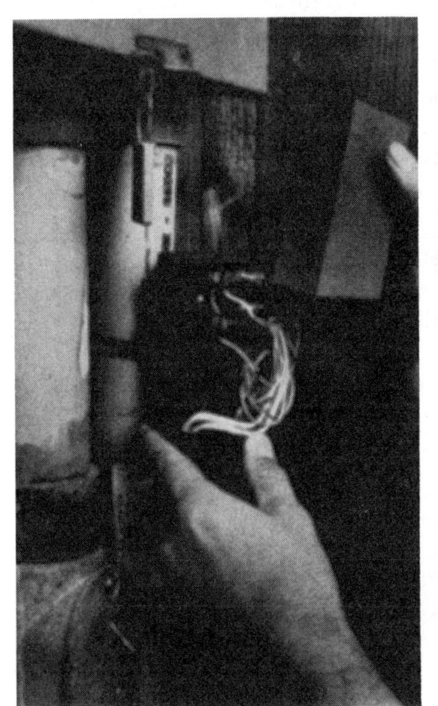

b. Multiple Protector Terminal Strip for Use in Apartments

Figure 1-7. Protectors

Whatever method is used to feed the cable from the protector to the inside of a home, apartment or small business, the telephone line pair usually connects to a terminal block called the *42A block*, a *modular interface*, or a *modular outlet*. All are shown on *Figure 1-8*. You may find the 42A block mounted in various places, on a baseboard, a wainscoat molding, a cabinet, a floor joist in the basement or ceiling joists in the attic. The modular outlet is normally mounted as a faceplate on an outlet box, which is just like a normal electrical outlet box.

In homes or small businesses where there are separate attics without a common passageway, there may be two lines coming into the building from the outside. Each cable pair for the inside starts at the protector and runs outside until a convenient place is found to go inside the building. Once inside, the cable pair usually ends in one of the three connectors shown in *Figure 1-8*.

COMMON WAYS OF INTERCONNECTION

The telephones that are installed in your building are always connected across the two wires coming from the protector, that is, all the telephones electrically are connected in parallel; however, the way they are placed on the line may be different. *Figure 1-9* shows three common ways. The 42A connector is used as a terminal strip to form the junction point of the incoming line from the protector and the fan out of the wire pairs to the installed telephones. Your telephone interconnection probably will correspond to one of these three. In *Figure 1-9a*, each wire pair connecting to each telephone starts at the 42A block. All connections originate at the 42A block. In *Figure 1-9b*, the telephone wire pair connections are continued from one telephone to another. The connections for the first telephone start at the 42A block. The wire pair for the second telephone is connected to the first telephone and the wire pair for the third telephone

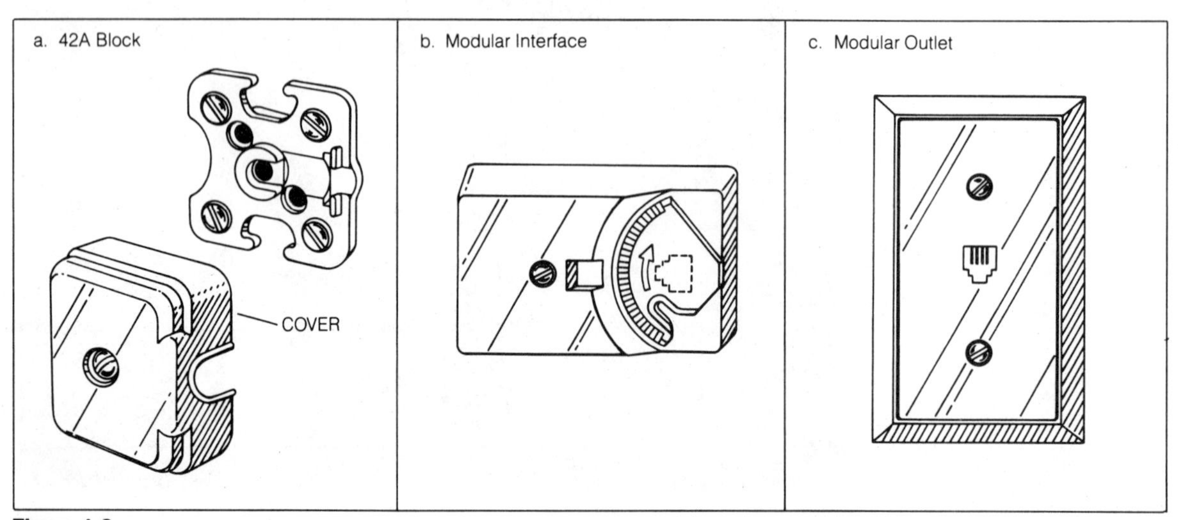

Figure 1-8.
Inside Connection of Your Telephone Line

is connected to the second telephone. The wire pair connection is extended as each telephone is added. Many times the bare wires making the connection at the first and second telephone are not separated at the terminals; they are just wrapped around the terminals or "looped" around the terminals. For this reason, this type interconnection is called a looped connection. In *Figure 1-9c*, the interconnection is a combination of the connections shown in *Figure 1-9a* and *Figure 1-9b*.

Now that we have some idea of the basic system, let's look at some of the major parts of a telephone and learn something about the parts and what they do.

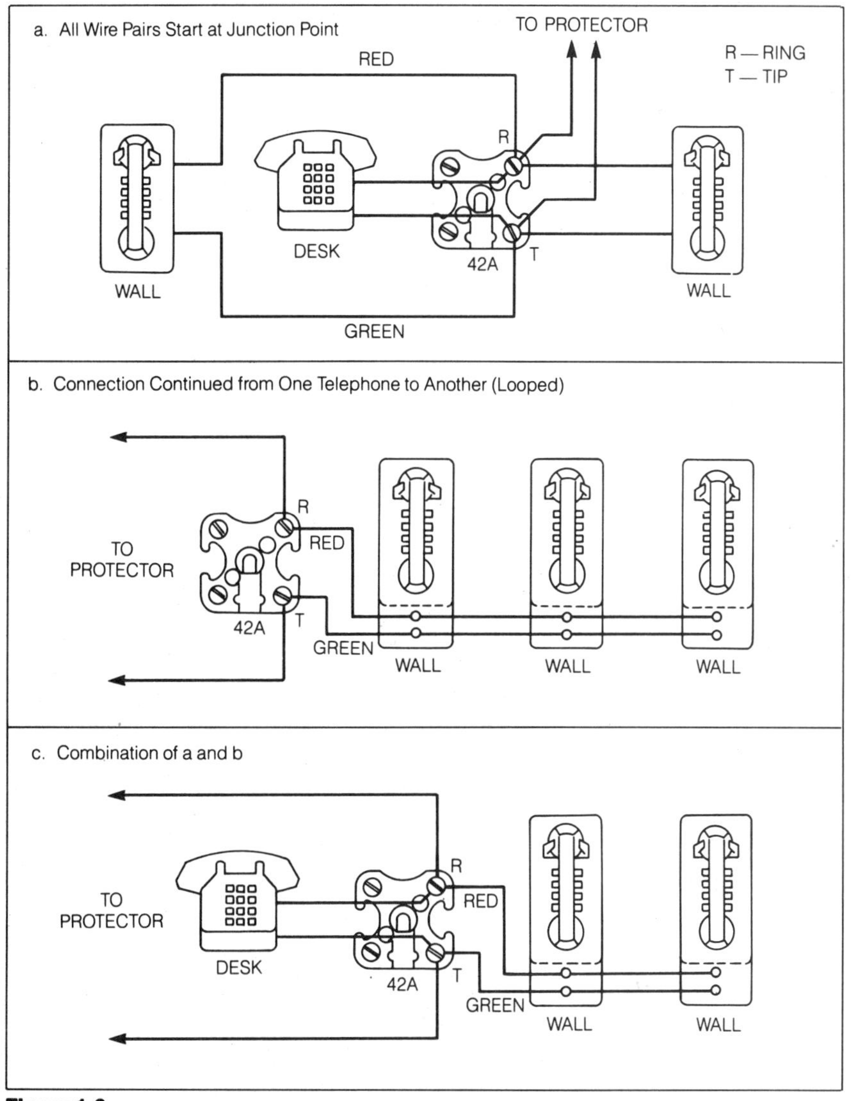

Figure 1-9.
Common Interconnection of Telephones

SWITCHHOOK

When your desk-mounted telephone has its handset in the cradle or when your wall-mounted telephone has its handset on the hanger, your telephone is said to be "on-hook". Lifting the handset from its cradle or hanger signals the central office that you want to use the system. This condition is said to be "off-hook". When your telephone is "off-hook", the telephone system sends you a dial tone when it is ready for you to use the system.

These terms "on-hook" and "off-hook" originated in the early days of the telephone, when the receiver was separate from the transmitter and hung on a hook when not in use.

When the handset is lifted to go off-hook it releases a spring loaded lever that closes switch contacts in the telephone. The spring loaded mechanism is called the switchhook. The buttons that pop up from a desk mounted telephone's cradle are the switchhook for that telephone set. Several examples of switchhooks are shown in *Figure 1-10*.

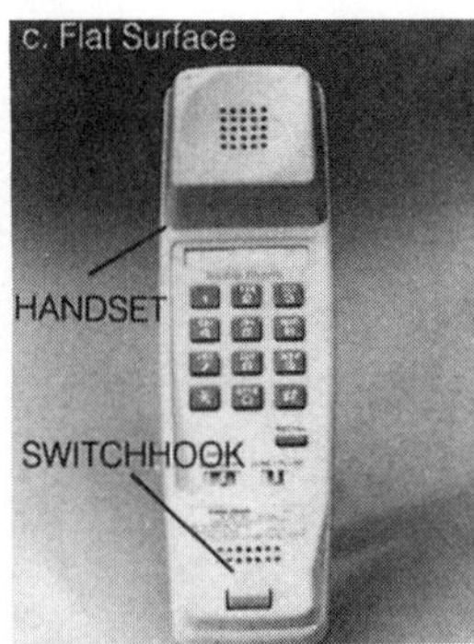

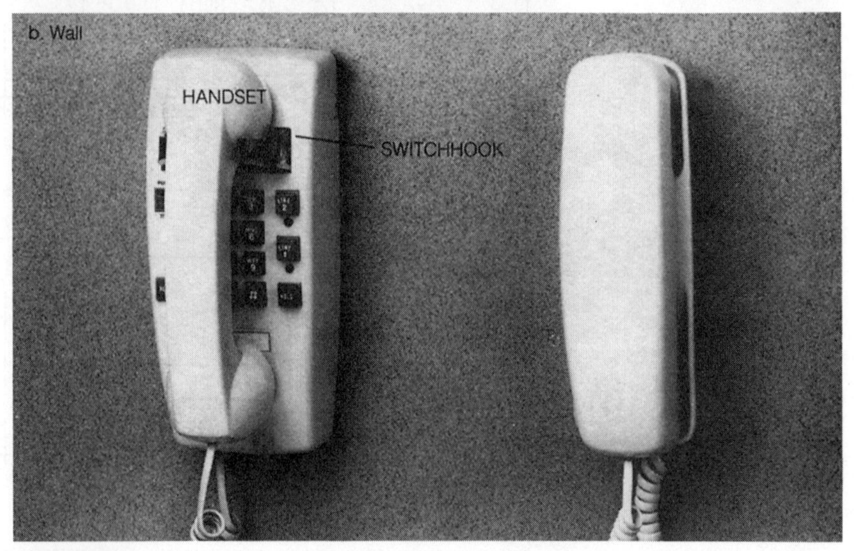

Figure 1-10.
Switchhooks
Courtesy of Radio Shack

HANDSET

Figure 1-10 also identifies the present day handsets. As stated previously, in the early days of the telephone, the receiver was separate from the transmitter. Such a telephone, introduced in 1914, is shown in *Figure 1-11*. The receiver was on a cord so it could be held to the ear as one talked into the transmitter. A typical handset of today has the receiver and transmitter included in the same housing as shown in *Figure 1-12*.

Transmitter

The transmitter shown in *Figure 1-12* is very much like one invented by Thomas A. Edison over 100 years ago. It is filled with carbon granules which have more or less pressure applied to them by a diaphram. The diaphram pressure is increased — and decreased — by the sound waves that hit the diaphram due to a person speaking. The changes in resistance cause changes in current in the transmitter circuit and therefore, changes in the local loop current to the central office. As a result, the sound waves of speech are converted to electrical signals which are detected and amplified at the central office and sent on to the called party. The called party's telephone transmitter is converting speech to electrical signals in the same fashion. Two wires are necessary to make the circuit connections to the transmitter.

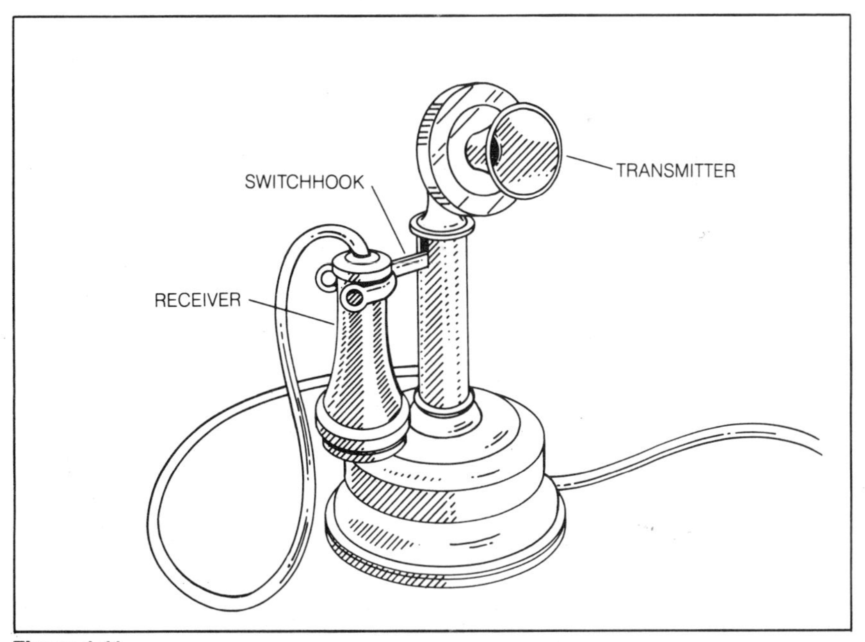

Figure 1-11.
Telephone Introduced in 1914

Receiver

The job of the receiver is to reverse the transmitter's function. It takes the electrical signals and reconverts them into sound. It is an electromagnetic device like a stereo speaker and works on the same principle. A coil of wire, wound with many turns of fine wire, is mounted in a strong magnetic field and is coupled to a diaphram or cone. Varying electrical current (caused by speech into the transmitter) through the coil causes the coil to move and with it the diaphram. The diaphram movement increases and decreases the air pressure in front of it and produces sound waves. The sound waves are picked up by the ear to complete the transfer of information from the person talking to the person listening. Two wires are also necessary to make the circuit connections to the receiver. Thus there are a total of four wires in the cable that connects the handset to the telephone set.

ROTARY DIAL

Almost everyone is familiar with the *rotary dial* on the front of a desk-mounted or wall-mounted telephone. *Figure 1-13* identifies the dial. When a finger is placed in the hole for the digit to be dialed, the dial is rotated clockwise to the stop. When the dial is released, it opens and closes switches coupled to the dial that break the circuit and then make it again. The breaking and making of the circuit causes the current in the circuit to stop flowing and then to flow again. As a result on-off current pulses occur

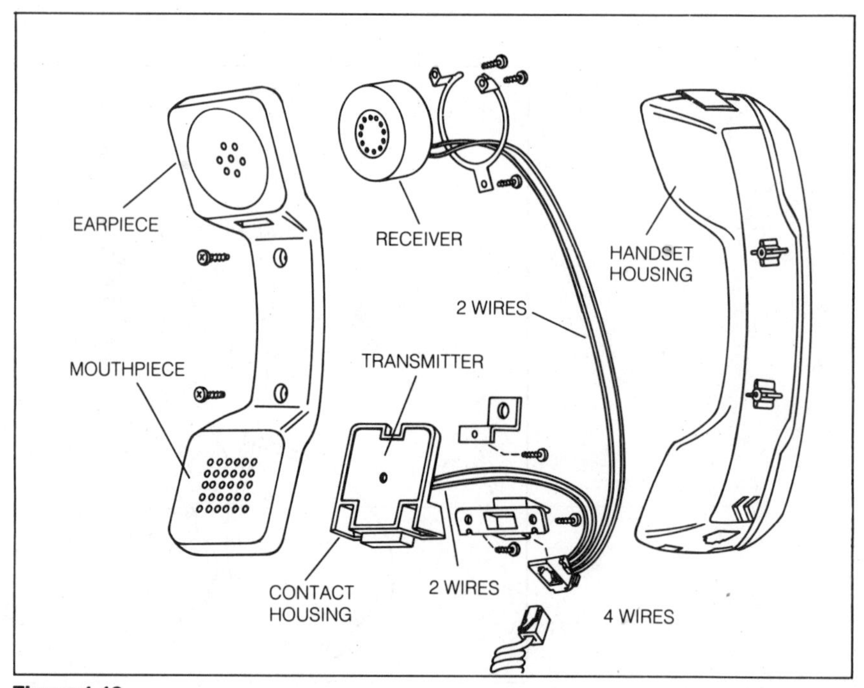

Figure 1-12.
Telephone Handset

in the local loop to the central office. The number of pulses correspond with the digit dialed. Circuits in the central office count the pulses and store the digit in switching circuits until all of the digits of the called telephone number are stored. When the last digit pulses are counted, the counting circuits in the central office set switching circuits based on the number received to make the connection to the called telephone. The connection may be to a telephone across the street or across the nation.

The dial is manufactured to produce 10 pulses per second as it rotates back to its initial position. Because the electrical signals are pulses, this type dialing also is called *pulse dialing*. In fact, electronic telephones like the one shown in *Figure 1-14* have no rotary dial but have push-button dialing. Electronically, they reproduce a set of current pulses at 10 pulses per second the same as the rotary dial. In their advertisements, manufacturers call these telephones pulse-dialing telephones that have push-button convenience. Many telephones also produce 20 pulses per second selectable between 10 and 20 by switch setting.

TONE DIALING

Since dialing provides the telephone system the address of the called number like addressing tells the post office where to send a letter, developments that reduce the dialing time and/or make the dialing more reliable are sought after by telephone companies. *Tone dialing* is such a development. It is much faster than pulse dialing — a 92% increase in time efficiency. As a result, less of the central office equipment is kept busy counting and storing digits until the complete telephone number is received. Therefore, the central office equipment can handle more calls.

Figure 1-13.
Rotary Dial Telephone

Figure 1-14.
Push-Button Pulse-Dialing Telephone
Courtesy of TeleConcepts, Inc.

Tone dialing is just what the term implies. Using an arrangement of push-button keys, combination audio tones with frequencies in the same range as voice frequencies are sent to the central office to identify the digits of the telephone number. *Figure 1-15* shows the push-buttons of a tone-dialing desk telephone. It has four rows and three columns of push buttons. The overlay over the keypad shows the frequencies that you hear in the receiver when a particular button is pressed.

Each time a button is pressed a combination of two tones is placed on the telephone line and sent to the central office. For example, if the "8" push button is pressed, a combination tone of 852 hertz (cycles per second) and 1336 hertz is sent to the central office. There are seven basic frequencies that form the 12 unique combination tones for the digits from 0 to 9 and the symbols * and #. The name given to the combination tone dialing is Dual-Tone MultiFrequency dialing or DTMF.

You must notify the local telephone company when you install a tone-dialing telephone because special circuits are required at the central office to detect the tone signals and identify the digits. They store the digits until the complete telephone number is received as with pulse dialing, and interconnect the correct lines to the called telephone based on the number received. But they do it much faster. Based on time calculations just to generate the signals, it takes a person, on the average, 11.3 seconds to dial a 10-digit number with a rotary dial. It takes an average of 1 second with push buttons — over ten times faster. Actual times may be a bit longer but the ratio remains the same.

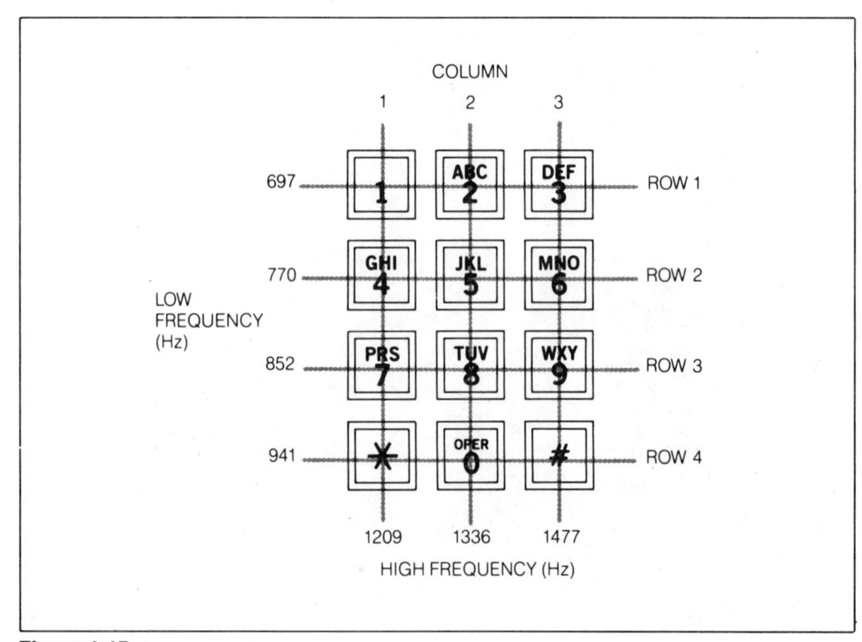

Figure 1-15.
DTMF Tone Dialing

RINGING — A CALL WAITING ALERT

The central office notifies you that a call is waiting by ringing your telephone. It alerts you that a caller has given your number (your address) to the system and the system has made the connection between the called telephone and the calling telephone. The central office places an alternating voltage signal of 90 VAC on the local loop to your telephone and this voltage excites the ringer in your telephone. The ringer is not easily seen because normally it is covered by the case of the telephone.

In *Figure 1-16* and *1-17*, the case of the telephone has been removed to show the inside of the telephone. *Figure 1-17* shows an old style telephone and old style ringer. The ringer is easily recognized by the two bells, a coil and a bell clapper. The signal from the central office is an alternating current signal that causes the clapper to be pulled back and forth by the coil to ring the bells. A mechanical adjustment is available to set the loudness of the ring. Such mechanical ringers are very much like the one invented over 100 years ago by Thomas A. Watson, Alexander Graham Bell's assistant.

Modern day telephones very seldom use mechanical ringers anymore. Instead they use an electronic amplifier and a speaker. *Figure 1-16* shows a speaker mounted on the telephone base underneath the keyboard. Some telephones may use a buzzer in an assembly similar to a bell, but the buzzer is still driven by an electronic amplifier.

The ringer always is across the incoming telephone line so that it can respond when the central office applies the signal. It draws no dc current from the central office line because it has a blocking capacitor in series coupling the signal to the ringing circuit.

SIDETONE AND THE SPEECH NETWORK

Whenever you hold the handset and carry on a conversation, you always hear yourself in the receiver as you talk. The small signal that is fed from the transmitter to the receiver is called *sidetone.* There is a circuit inside the telephone, again one that normally is not seen because its covered by the case, that sets the amount of signal that is used as sidetone. The circuit has a number of names. Very early and still today it is called the induction coil or hybrid network. With the advent of electronic telephones it is being called the speech network. It serves two purposes. The first is to couple the talking signals of the transmitter with its two-wire connection to the two-wire telephone line with only the amount of sidetone desired. The second is to couple the signal received on the two-wire telephone line from the calling party to the receiver over its two-wire connection and not couple any of the signal to the transmitter. So one can think of the speech network as a matching or balancing circuit that couples the conversation forward from the transmitter to the telephone line with a small amount of the signal sent to the receiver, and back from the telephone line to the receiver without a signal to the transmitter.

Figure 1-17 shows a common version of the speech network, induction coil, or hybrid network. In the older style telephones, it is called a potted circuit because all the components are sealed together with a potting compound and cannot be seen. The only things exposed are the screw terminals for connection. In the newer telephones, the components may be exposed and the connections made to a circuit board. You will only come across this network in your installations when you are changing old style cords to modular cords. The cords may run from headset to the telephone set, or from the incoming telephone line to the telephone set. The incoming old style cord connections normally are made to the speech network. In more modern telephones, all cord connections are made through modular connectors.

TYPES OF TELEPHONES

Throughout this book we will be describing the installation of basically three types of telephones, desk or table telephones, wall-mounted telephones and extension telephones. *Figure 1-10* shows a variety of each of these. You may already be using several of these types.

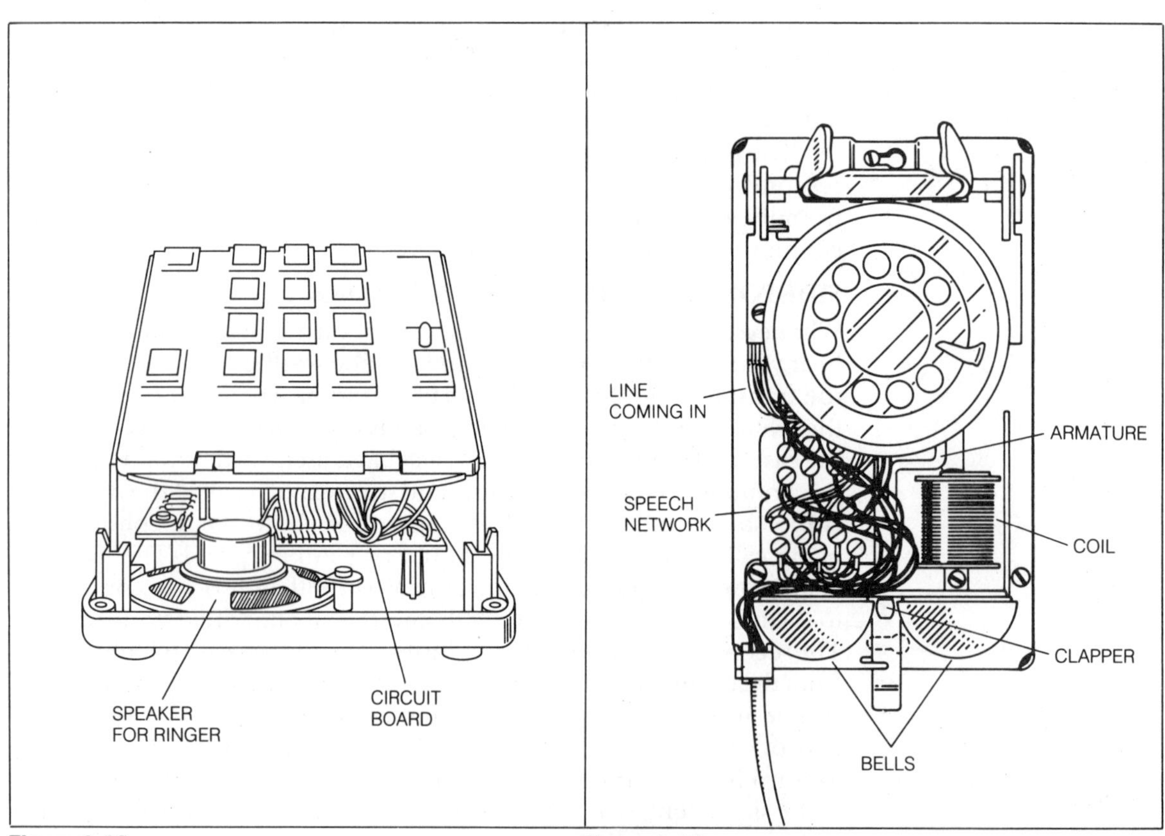

Figure 1-16.
Electronic Ringer

Figure 1-17.
Old Style Speech Network and Bell Ringer

TYPES OF TELEPHONE SET CONNECTIONS

Throughout the book we will be talking about the two common ways of connecting the telephone set to the telephone line. What is classified as the old style connection has telephone set cords coming directly out of an outlet faceplate through a small hole, or directly from a 42A block or connected by a 4-prong plug into a 4 -prong jack in an outlet faceplate. These are shown in *Figure 1-18a*. Contrasted to the old style is the new style modular connector system represented by the plug and jack of *Figure 1-18b*. Your interconnections to your telephone sets will be of one or a mixture of these two classifications.

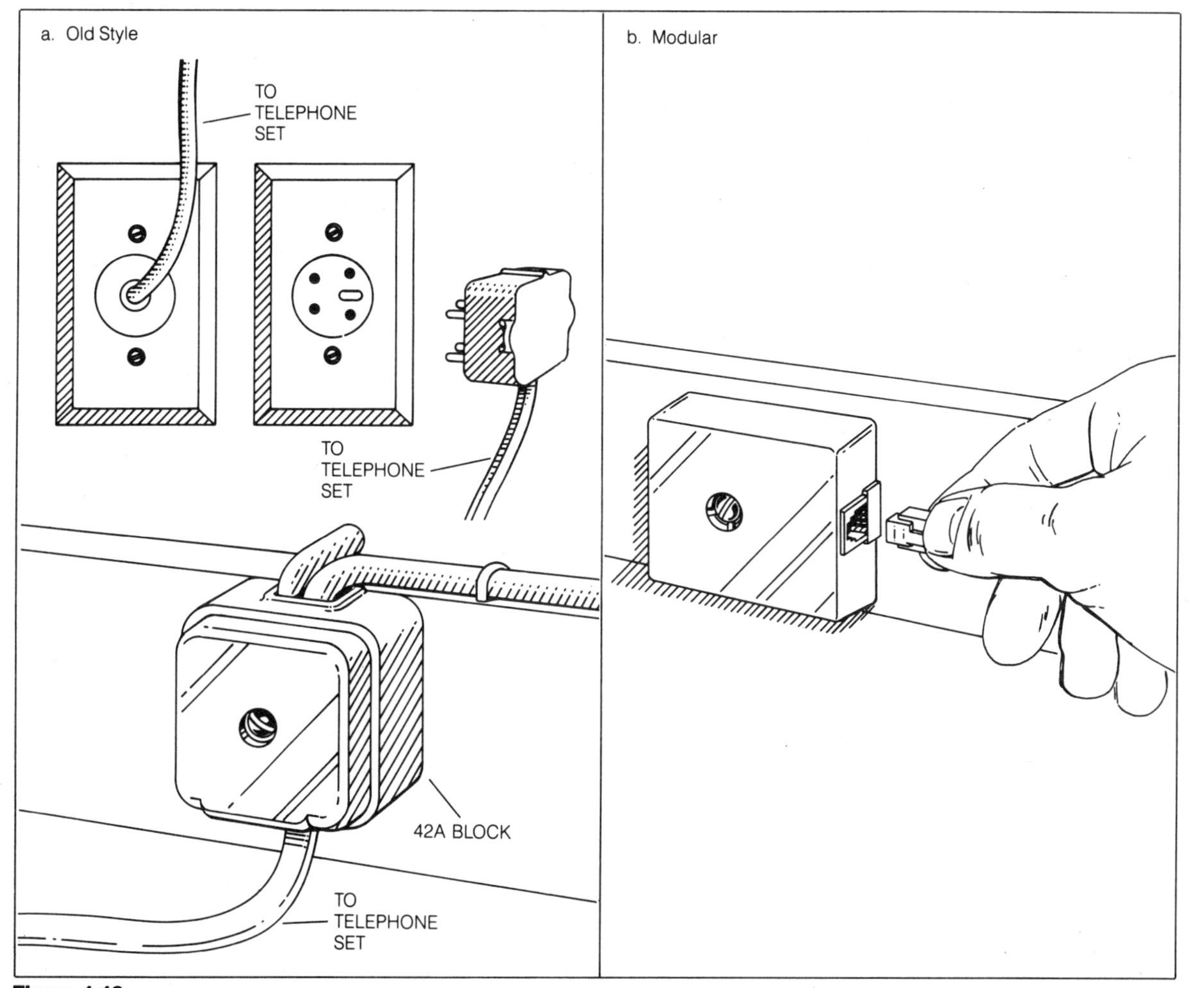

Figure 1-18.
Two Basic Connection Systems

WHAT'S TO COME

With the basic understanding obtained from this chapter, the installation instructions begin with the next chapter. First, home and apartment installations, then business, then pre-wiring, and finally accessories. Not all installations are correct the first time; therefore, a chapter on troubleshooting is included, which also will be of help in maintenance and repair.

Throughout the book we will be dealing with installations that take the minimum of time, use readily available parts and are accomplished in a straight forward fashion with step-by-step instructions that are well illustrated.

ABOUT THE NEWER TELEPHONES

Many new telephones are coming on the market with a great deal of electronics inside. They may use small ceramic disk microphones and electronic pulse circuits for pulse dialing. They may have different kinds of speech networks inside and use multitone generators for the ringer. They will all use modular plugs and jacks for connections. However, their functional operation is the same as has been described. They interface with the telephone system in the same manner and use the same wire pair connections. Many of them will be powered from the central office the same as the older style telephones. They are likely to be light weight; use semiconductor integrated circuits for their electronics; use much less power; be very reliable and have at least some, if not many, added features over the standard telephone set; and for a lower cost.

ABOUT PURCHASING TELEPHONES

Telephones purchased from any store or other conveniently available outlets are yours completely, the case, the electronics, push-buttons, switches and all parts inside.

In the past, many people have thought they have purchased telephones from the telephone company when actually they have purchased just the outside case. The inside mechanisms and parts still were the property of the telephone company. Make certain that you are buying the complete telephone.

ABOUT INSTALLATIONS

Telephone service is brought into your building from the telephone company's protector to an initial termination. It is from this initial termination that your installation should take place. If you need to make an additional connection to the protector, check with your local telephone company to get their approval for you to make the connection. You cannot make your own telephone installation if you are on a party line or on a line with a coin operated telephone.

ONE THING MORE

Rotary phone systems still enjoy a great deal of use both inside and outside the U.S.A. Most countries continue to use predominantly rotary phone systems. In the United States, touch-tone digital phone equipment has soared in popularity during the past fifteen years due to advances in digital and computer technology, but it has not yet completely replaced the old-style rotary system. As late as 1990, in fact, AT&T introduced the first answering machines to respond to rotary phones and voice responses, so customers did not necessarily have to dial from touch-tone phones to call home for their answering machine messages. Part of the reason behind the continued use of rotary phones is due to their cost. To lease a rotary phone from a phone company costs approximately 20% less than it costs to lease a touch-tone phone. Also, phone companies tend to charge more for general touch-tone phone usage than for rotary. In addition, many older homes and buildings in the United States still have their original rotary installations because the individuals who own these structures don't believe it to be cost-effective to convert over to touch-tone systems. This book contains information to help rotary phone users add to or maintain their systems, or convert to touch-tone systems. If you are someone who wants to convert from a rotary to a touch-tone system, we hope to show you how to do the conversion yourself and save a great deal of money in the long run.

VOICE
DATA

2 ASSESSING WHAT YOU HAVE

Chapter Contents	*Page*
Locating Outside Line and Protector	*2*
The Ground Line	*5*
Initial Input Termination	*6*
42A Block	*6*
Modular Interface	*7*
Attic	*7*
Basement	*8*
Home and Apartment Distribution	*8*
Cable Distribution	*11*
Completing the Assessment	*12*

Before actually taking your tools and beginning an installation, it is best to look over the proposed installation and do a bit of planning. This gives you an opportunity to select the materials and the correct tools that are needed, and will save time, effort and a lot of false starts.

After reading this chapter you should:

A. Know if your telephone system is modular or old style.
B. Have an idea how it is interconnected.
C. Be able to sketch a simple floor plan showing location of telephones.

The planning begins by looking over (assessing) your present telephone installation.

HOW ARE HOME OR SMALL APARTMENT TELEPHONES INSTALLED?

Your assessment starts by locating where the telephone service enters your building from the outside.

Locating Outside Line And Protector

Go outside and determine how the telephone company line runs to your building. Your key element is the protector. Newer styles are shown in *Figure 2-1.* They may have a plastic cover or a metal cover. They have a cable (usually black) containing at least four wires (four conductors) which is the line from the local telephone company, connected as shown in *Figure 2-2.* Also connected, usually with smaller wire, is a four conductor cable that is the telephone line into the building. Sometimes several cables are connected from the protector and run into the building.

A ground wire may run to a conduit or water pipe or copper pipe for grounding or the protector may be strapped to one of these to make the ground connection. The cable from the telephone company may come to the protector from overhead or from underground. *Figure 2-3* shows typical examples. Recall that the protector serves as a lighting arrester and that you are not to change the connections of the line coming from the telephone company. Your installation possibly may occur after the protector, never before the protector.

a. Plastic Cover b. Plastic Box

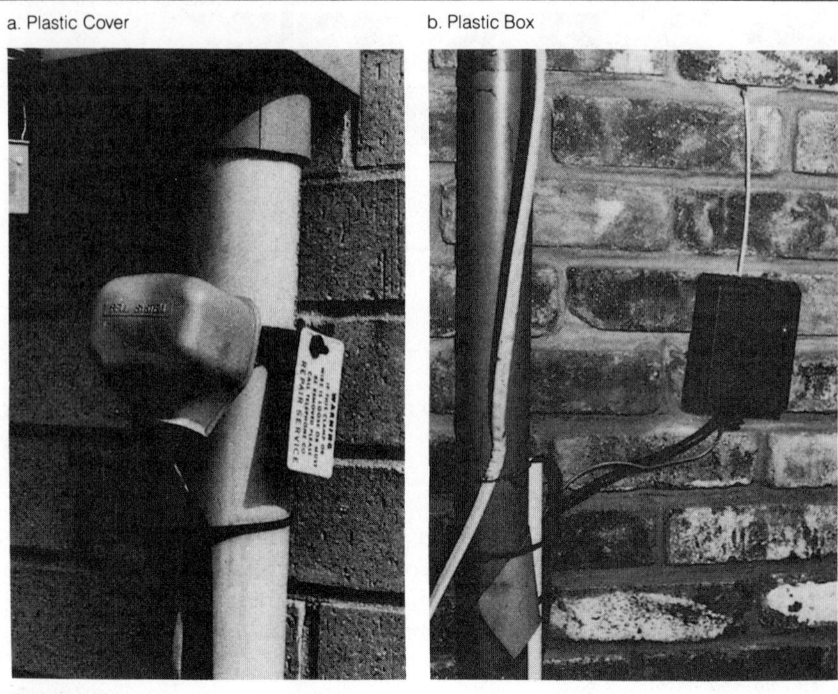

Figure 2-1.
Newer Style Protectors

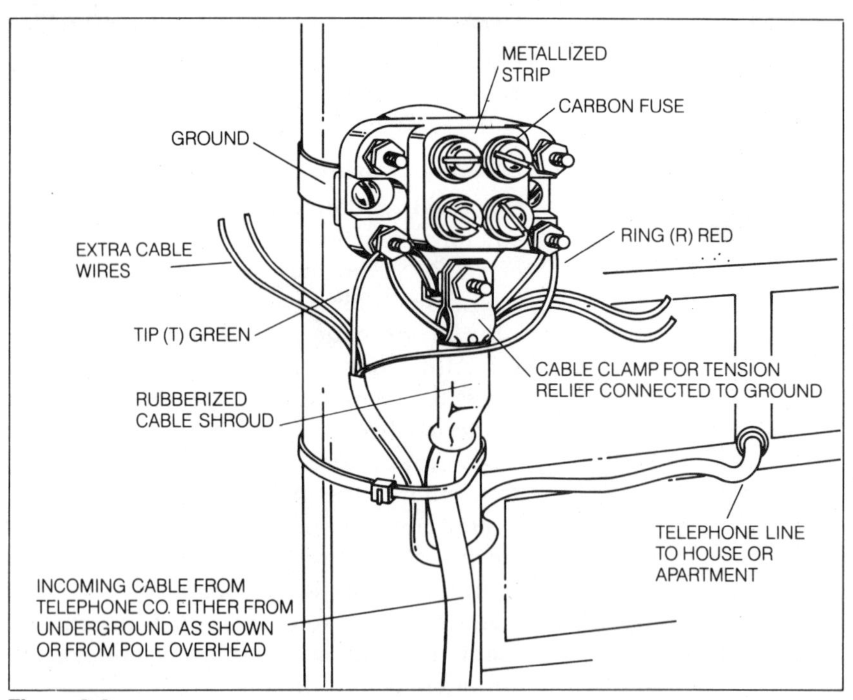

Figure 2-2.
Protector (Plastic Cover 3″ x 2½″ x 2″) Connection Details

Note in particular, how and where the telephone line enters the building. In *Figure 2-3a* the cable carrying the telephone line into the house goes directly into the attic at the protector. In addition, as shown in *Figure 2-3a,* another cable comes from the protector, runs down the side of the house, and goes directly through the brick wall into the house. A similar case is shown for the conduit mounted protector of *Figure 2-3b.* A hole has been drilled directly through the outside brick wall and the cable inserted and pushed through to provide a telephone line into the house at this location.

Older Style Protectors

In older homes and small apartment buildings, the telephone company line may come overhead or underground and go directly through the building wall as shown in *Figure 2-4a.* There is no protector on the outside, but the cable wires are connected to a protector that is inside the building. *Figure 2-4b* shows a very early style protector mounted to the floor joists in the basement with the incoming and outgoing telephone lines connected.

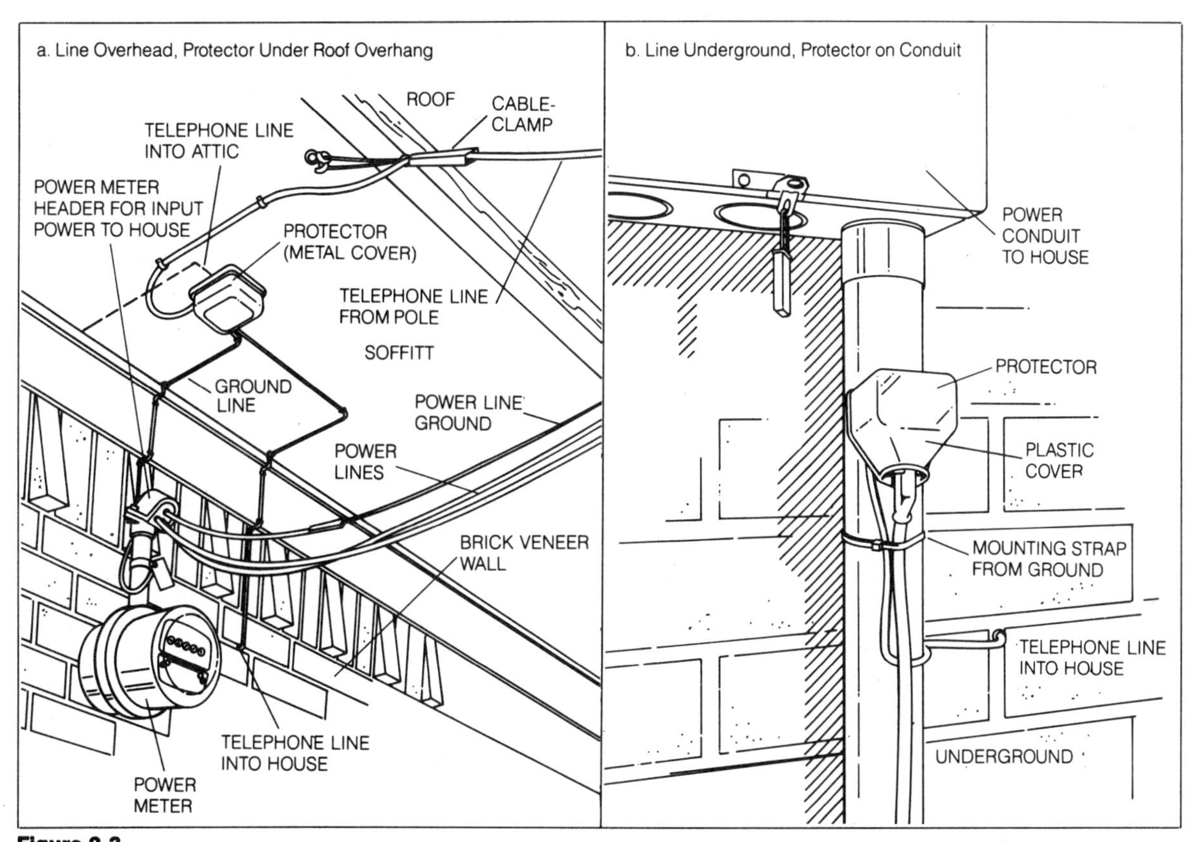

Figure 2-3.
Telephone Lines to Protector

LARGE MULTIPLE-DWELLING BUILDING INSTALLATIONS

If you are living in a large apartment building, a condominium, or large townhouse complex, you will not be concerned as much about where the company telephone lines enter the building and how they are distributed to your apartment. Your concern is — where does the telephone line enter my living space? As discussed in Chapter 1, the telephone lines from a multiple-unit protector commonly end up in the apartment terminated by a 42A block or, with the new modular system, by a modular interface block or a 4-terminal outlet box.

THE GROUND LINE

In order for the protector to provide the protection against high-voltage spikes on the line due to lightning, it is very important that the ground connection be a very solid substantial connection. When the protector is mounted directly to a conduit, water pipe, or a copper bar, the ground connection is made through the strap that holds the protector in place. When the protector is mounted to a wood or brick surface, a separate ground wire is connected to a conduit or water pipe or run to a copper bar driven into the ground. *It is extremely important that these connections remain secure and in good working order.* If you disturb a ground connection in your installation, make sure that it is put back in place securely.

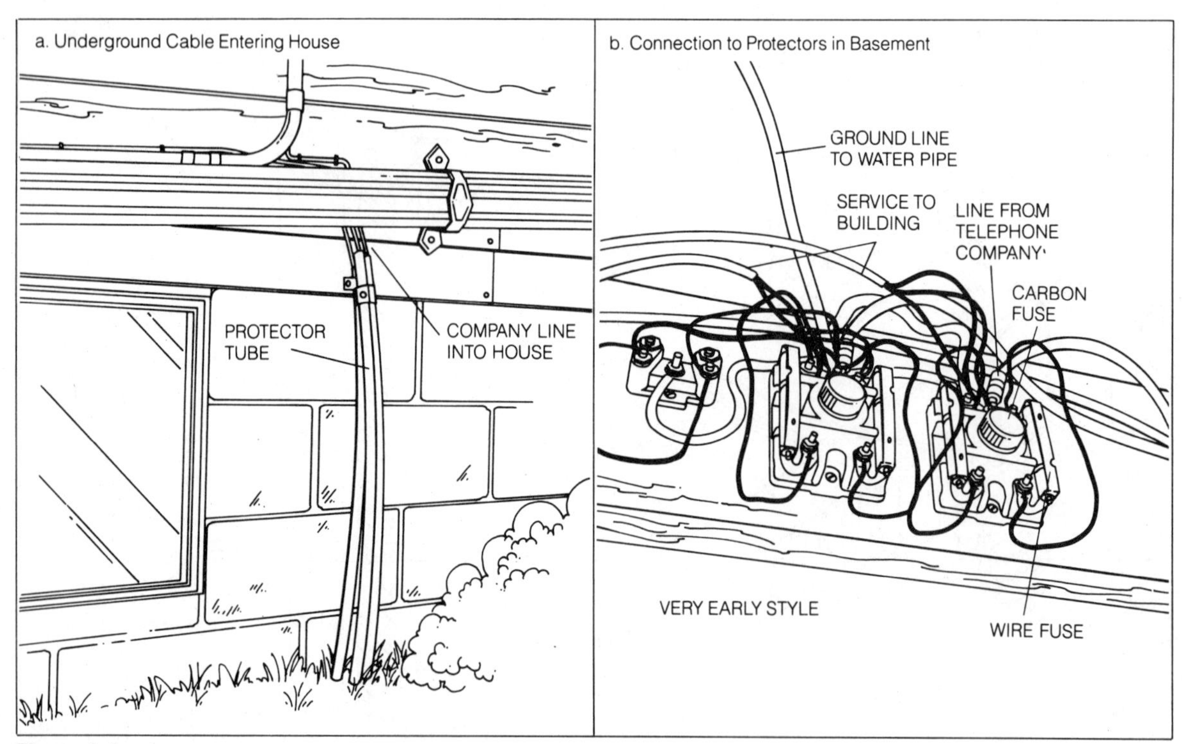

Figure 2-4.
Old Style Protector

INITIAL INPUT TERMINATION

Now that the protector and the place where the telephone line enters the building have been located, go inside and locate where the telephone line terminates as it comes through the wall, or into the attic, or into the basement or from underneath the floor. The most common way in mobile homes is for the line to come in under and through the floor.

42A Block

As mentioned previously, a very common place for the telephone line to connect initially as it comes in from the outside is at a 42A terminal block. *Figure 2-5* shows several ways that the 42A block with its cover might be mounted. If the building was not prewired with telephone wires before the interior walls were finished, then *Figure 2-6a* is very common. If the building was prewired then *Figure 2-6b* might be a common termination. This is especially so if a desk-mounted telephone is nearby. The cover plate with a hole in the middle for the telephone cord is a very neat trim for the outlet box. In some cases this cover plate is in place and no telephone has been installed. The 42A block inside the outlet box still forms the initial terminating point for the telephone line from the outside.

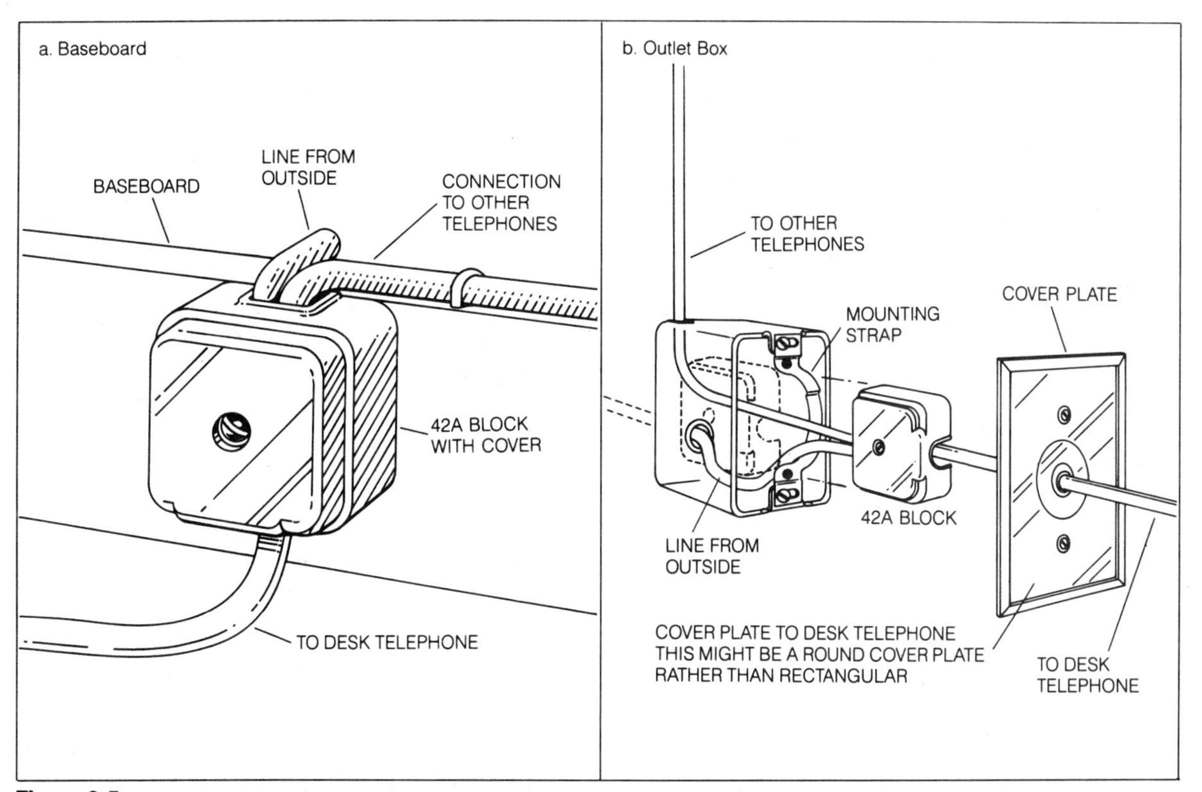

Figure 2-5.
42A Block Terminations

Modular Interface

For the newer modular connector systems, the initial connection of the telephone line inside the house is made one of the several ways shown in *Figure 2-6*. If the incoming line comes into an outlet box, *Figure 2-6b* is the likely connection; if not, then *Figure 2-6c* is common. In many modern home installations, there may be a junction box near the power switch panel that contains the connection of lines that run to each outlet box from this junction point. This is shown in *Figure 2-6a*. A modular plug from the junction point connects to the modular jack of a standard network interface jack (RJ11W or RJ11C). In new modular installations your installations *must* be through this modular interface.

Attic

If the telephone line from the outside enters the attic, there is a good chance that the 42A block is used as a junction terminal strip. If it is not, then the line has been run from the attic to a junction box in a wall, looped

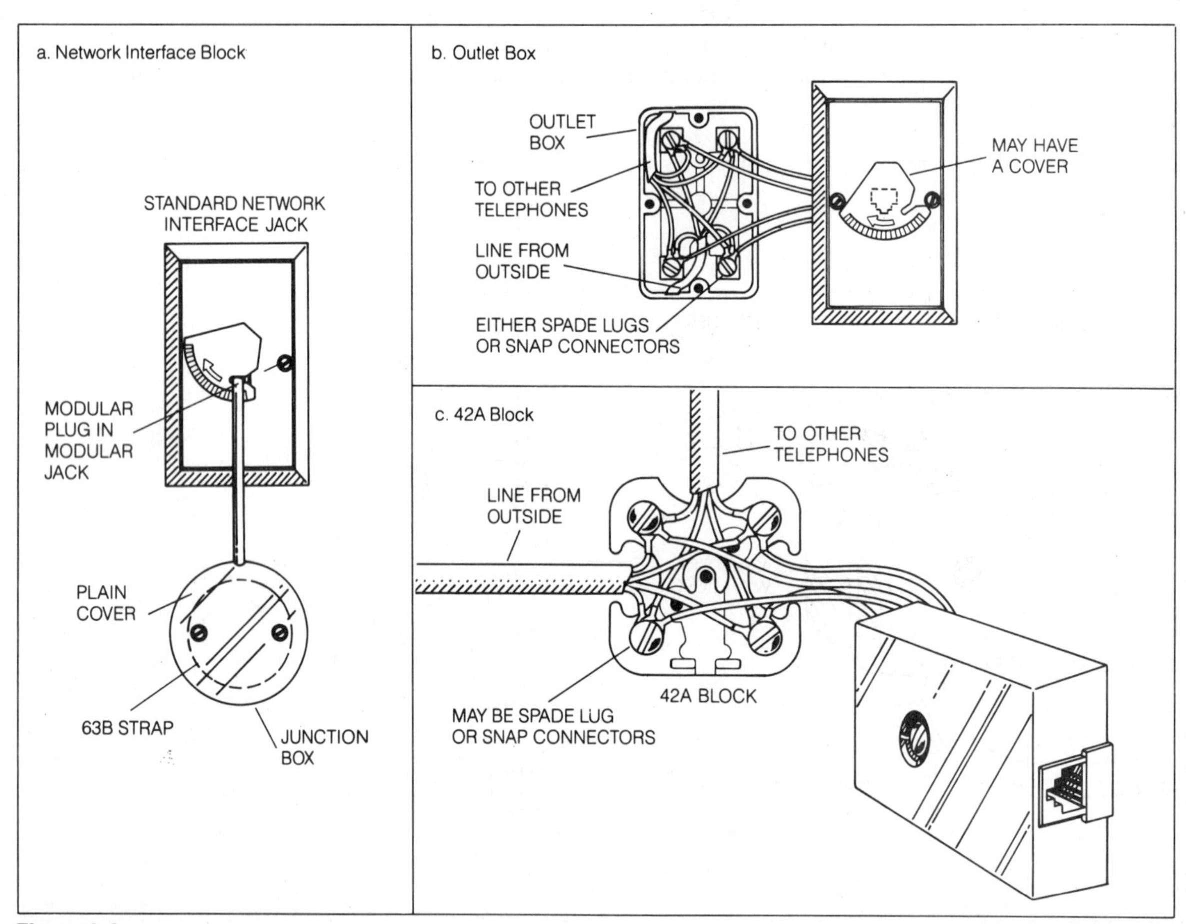

Figure 2-6.
Initial Inside Connections for Modular System

back up the wall into the attic and down again to another junction box until all outlets are wired (See Chapter 9 for reference).

Basement

If the telephone company line does not run to a protector on the outside of the building, then it comes through the outside wall and its initial termination is a protector as shown in *Figure 2-4*. Present day installations would use the new style protectors. From the protector, a 42A block may be used as a junction terminal strip just as in the attic. From this point, if the building has been prewired, the cables are distributed to the wall locations and up into the walls to outlets. For installations that are not prewired, the cables go up through the floor to 42A blocks at the respective telephone locations. The 42A blocks are mounted outside the walls on baseboards, on wainscoat trim, or in outlet boxes.

IN HOME OR APARTMENT DISTRIBUTION

The next step in the assessment is to determine the way the existing telephones are connected to the line.

Old-Style Jacks and Plugs

Figure 2-7 shows the various types of old style jacks and plugs that were used to connect desk, table and extension telephones. If the home or apartment was wired for telephones before 1973 most likely the old style connectors were used. If the building was prewired then the jacks probably are mounted in outlet boxes. If installations were made after the

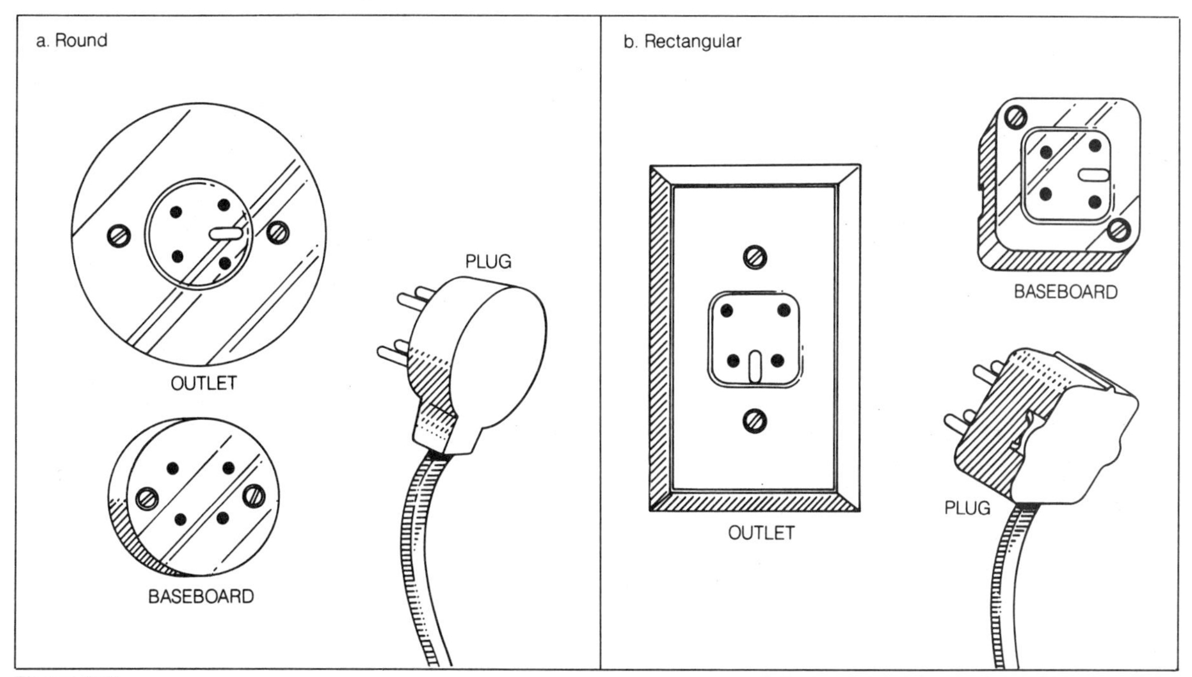

Figure 2-7.
Old Style Telephone Jacks and Plugs

building was finished, then the jacks are likely to be mounted on the baseboards.

As shown in *Figure 2-8,* any wall mounted telephone has the wire pair connected directly to the speech network in the telephone. There is no extra connector.

Modular Connections

All new phone equipment now is manufactured with modular connectors. The modular connectors consist of modular jacks and a modular plug.

Three of the most common types of modular connections are for wall-mounted telephones, table-mounted telephones and extension telephones. Unlike the old style jacks and plugs, when modular connections are used usually there is a connector on the wall-mounted telephone. *Figure 2-9* shows the common type of modular connections for wall-mounted telephones. The modular plug in this case is mounted directly to the telephone and inserts into the wall-mounted jack (wall telephone modular jack), as the telephone base is attached by engaging the slotted catches over the mounting pins on the wall-mounted modular plate.

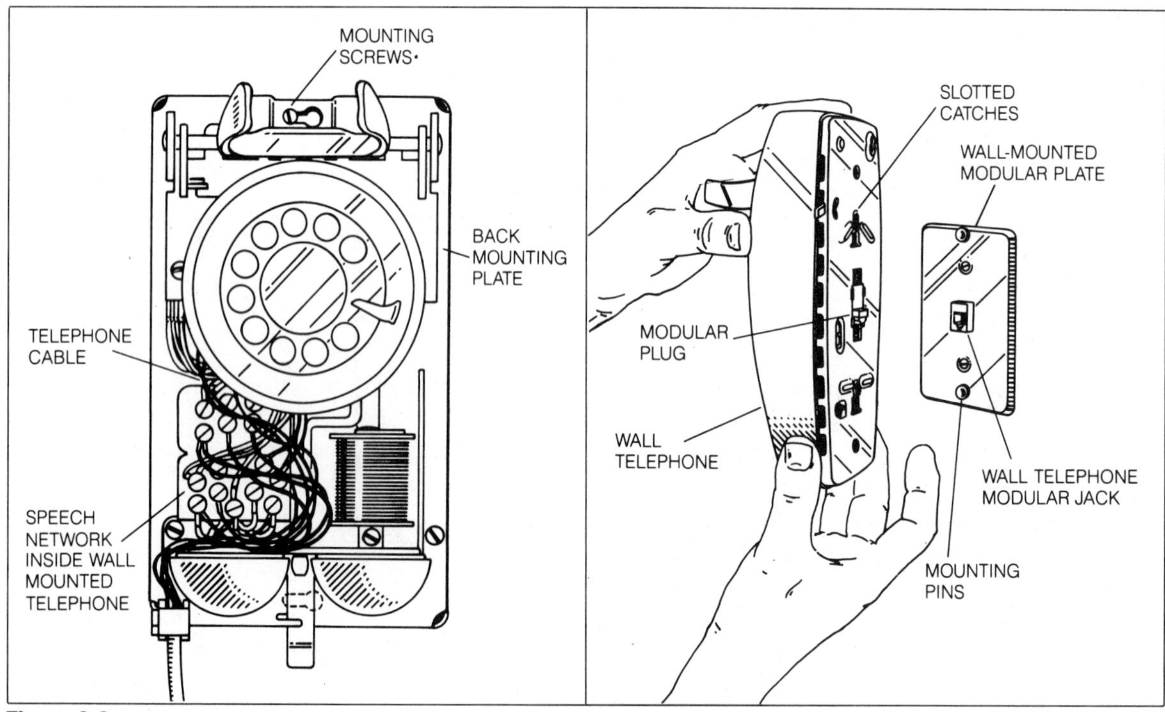

Figure 2-8.
Wall-Mounted Old Style Telephone

Figure 2-9.
Wall-Mounted Modular Telephone

For desk, table and extension telephones the usual type of modular connectors are shown in *Figure 2-10*. Rectangular face plate mounted jacks that fit standard electrical outlet boxes are shown in *Figure 2-10a*. The round outlet fits a special box or can be mounted directly to the wall. The telephone cord with its modular plug plugs directly into the jack.

If extension phones are the choice, the modular connector shown in *Figure 2-10b* can be placed right over a 42A block. The remaining jack, shown in *Figure 2-10c* usually is surface mounted to a baseboard. A long flexible telephone cord with its modular plug completes the connection to any of these modular jacks.

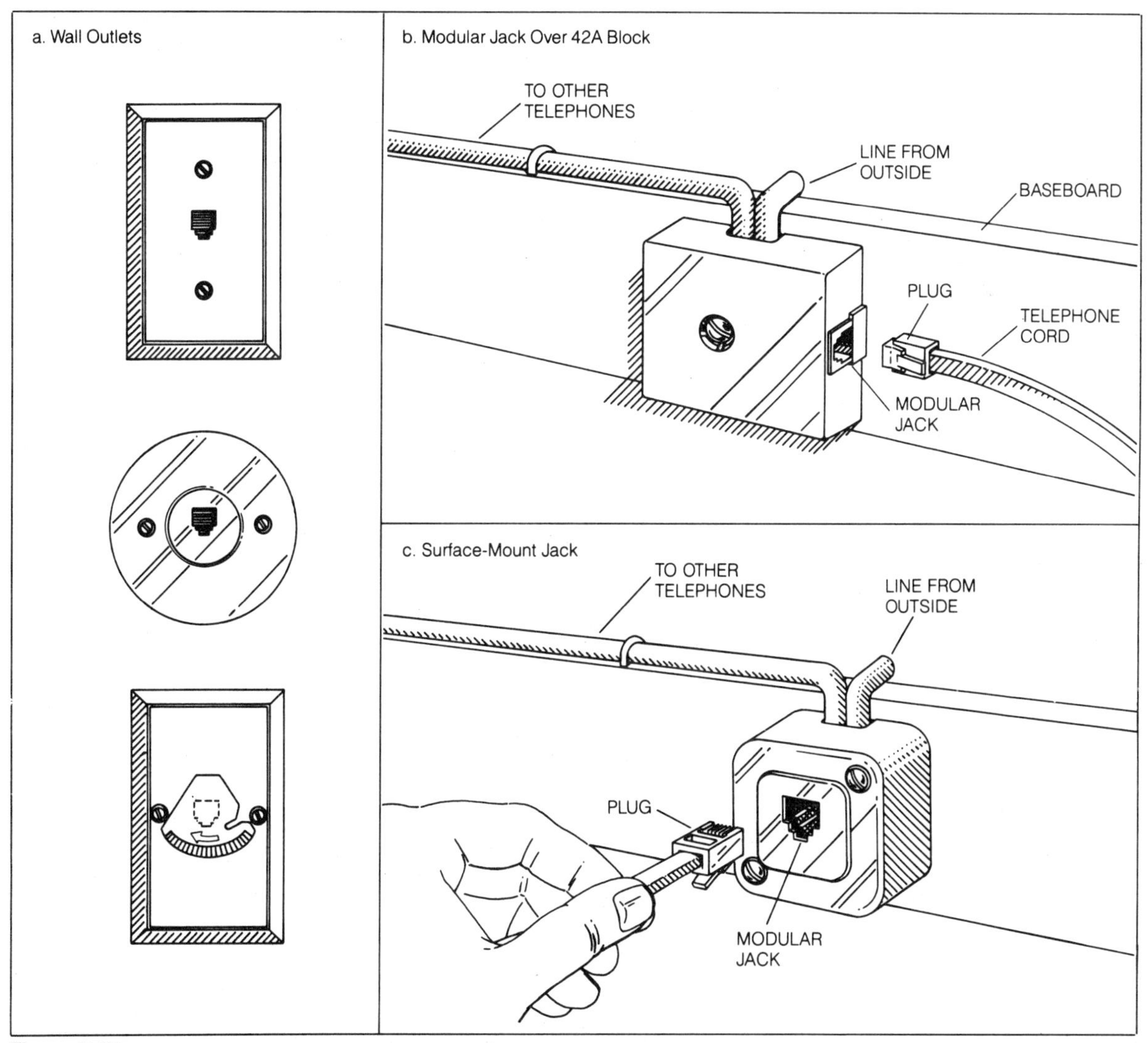

Figure 2-10.
Modular Desk, Table and Extension Connectors

CABLE (STATION LINE) DISTRIBUTION

Having located the initial terminations and all the jacks, plugs and direct connections of the telephones to the line, the remaining task is to determine how the cables run between the outlets. *Figure 2-11* illustrates one way that the telephone lines may have been wired. Any one of the connections shown in *Figure 1-9* may have been used, especially the looped connection if the building was prewired. If there are wires running along baseboards, in closets, or in cabinets, then probably no set pattern of wiring has been maintained. Under these circumstances, it should be easy to trace the interconnections. You may have to remove some of the outlet plates and covers to determine the number of cables present and the direction of the cable run. In many cases, the cable wires will not be broken but looped around each screw terminal as shown in the expanded view of *Figure 2-11*. However, within a short period of time you should be able to draw a simple diagram like that shown in *Figure 2-11* for your specific home or apartment.

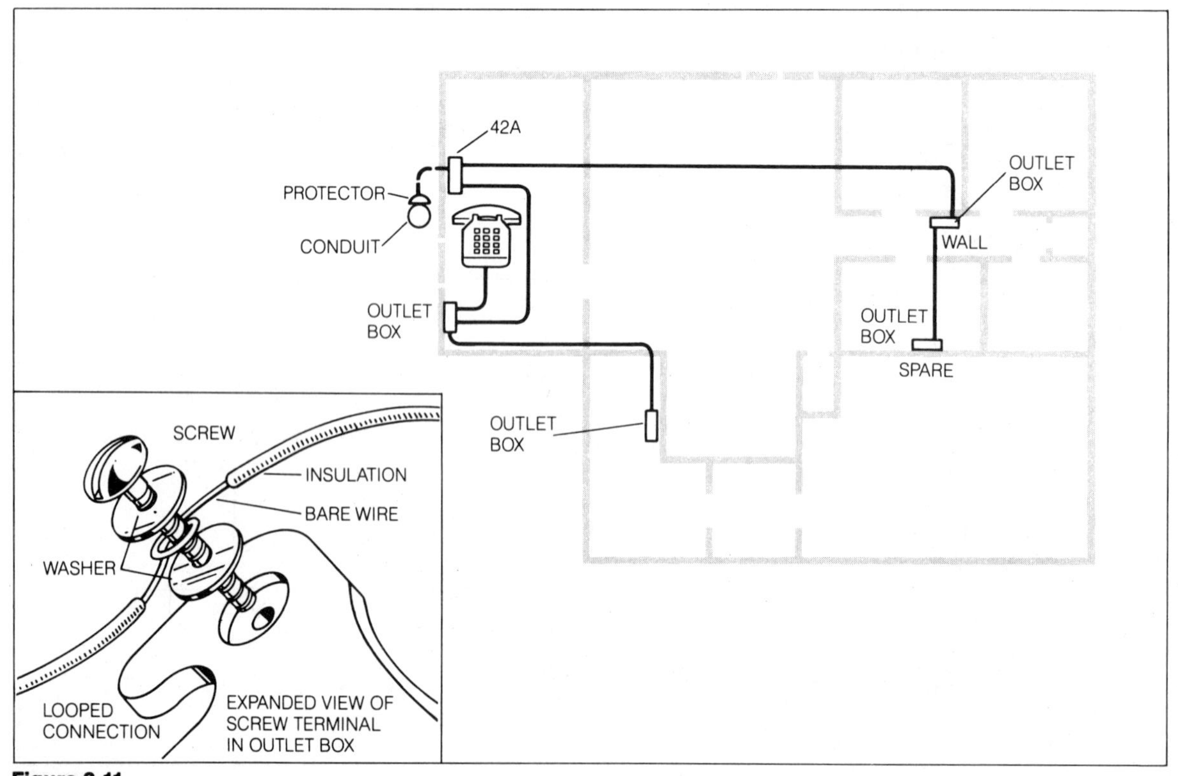

Figure 2-11.
Station Line Interconnection

PULSE OR TONE DIALING?

If all of your telephones are rotary dial telephones, your present telephone service is likely pulse-dialing service. If you have push-button telephones and they don't have PULSE-TONE selection switches, then they likely are pulse-dialing telephones and you are operating with pulse-dialing service.

Here's an easy way to determine if your present telephone service is tone-dialing service. Using a telephone that has a PULSE-TONE selection switch, set the switch on TONE. When you lift the handset off hook, you should have a dial tone. Dial any number from 1 through 9 except 0. If you no longer have a dial tone, you have tone-dialing service. If the dial tone has not stopped, you have pulse-dialing service, and you must have your PULSE-TONE switch on PULSE to operate properly.

You will have to call your telephone company and request tone-dialing service to operate using tone-dialing telephones.

COMPLETING THE ASSESSMENT

With all the data in hand, sketch out a floor plan of the building. Show the location of the existing telephones and the interconnection of the station lines. It should look similar to *Figure 2-11*. After completing the floor plan, follow these steps:

1. On the floor plan make a preliminary decision that locates each of your telephones that you want to install. This plan shows the existing telephone company telephones that will be replaced or moved and the new telephones that will be added, if any.
2. Decide if the telephone at each location is to be a wall-mounted, desk, table, or an extension telephone. In most cases, the existing installation will dictate the type.
3. Read the installation instructions. They are divided according to modular and old style systems, whether or not you are replacing or adding telephones and the type telephone that is being installed.
4. Modify the preliminary plan if necessary.
5. Make a list of materials (called bill of materials) so that the parts that must be purchased are identified (See *Figure 9-3*, Chapter 9).
6. Check the tool list to make sure necessary tools are available.
7. Purchase necessary parts.
8. Complete the installations.

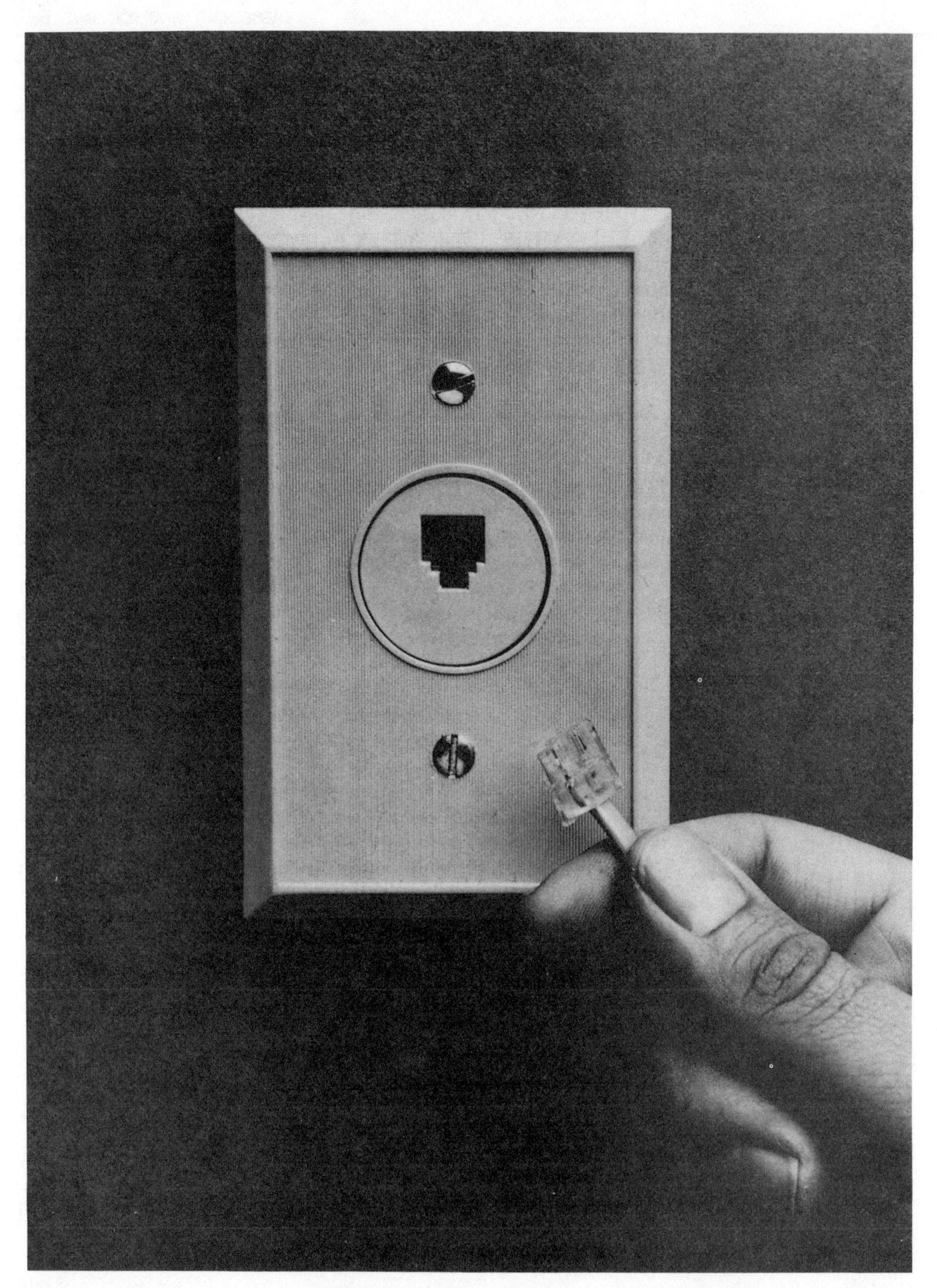

3-1

3 REPLACING MODULAR TELEPHONES

Chapter Contents — *Page*

General 2
FCC Registration 2
REN Ratings 3
Tone Dialing Lines 3
Maintenance 4
Notes of Caution 4
Tools 5
Replacing Extensions 7
Replacing Desk or Table Telephones 8
Replacing Wall Telephones 10

This is the first of the chapters that contains step-by-step instructions for telephone installations. It explains how to replace present rental telephones in a modular system with telephones you have purchased. If you are replacing telephones in an old style system turn to Chapter 4. If you are adding telephones, turn to Chapter 5 to find out how to run the cables and Chapter 6 and 7 for the installation instructions. A general section in this chapter applies to all installation chapters.

After reading this chapter you should:

A. Know how to report the FCC registration numbers of the telephones that you buy to the local telephone company.
B. Be aware of the *"NOTES OF CAUTION"*.
C. Know which tools are required before starting an installation.
D. Be able to replace rental extension, desk and wall-mounted modular telephones with telephones you buy.

GENERAL

FCC Registration

Any telephone that you purchase and install must have the following information so that you can notify your local telephone company:

Complies with Part 68, FCC Rules
FCC-Registration No. ______________
Ringer Equivalence ______________

In addition the following information is usually listed:

Manufacturer ______________
Model ______________
U.S.O.C. Number ______________
Serial Number ______________
Date of Manufacture ______________

The serial number may be in a manufacturer's selected location and the date of manufacture may be in some special code. The U.S.O.C.[1] number is the way the modular connector is wired to make the connection.

[1]U.S.O.C. — Universal Service Order Code

REN Rating

Your local telephone company only guarantees to supply a finite amount of ringing power — usually for five telephones. Each standard ringing power unit is a REN. If the telephones that you install add to more than 5 REN, then the amount guaranteed by the telephone company has been exceeded, and your telephone may not ring. It will depend on how far you are from the central office. Standard telephones take 1 REN to ring. Many electronic telephones take much less than 1 REN so more of them can be connected to the line.

There is also a letter after the REN number. It tells the ringer frequency to which the telephone will respond. The letter A stands for the normal single-party ringing frequency of 20 ± 3Hz and 30 ± 3Hz. These are the ringer frequencies normally supplied by the local telephone company. The letter B on the REN number means that the telephone will respond to frequencies from 15.3 Hz to 68.0 Hz, a much broader frequency range. It will work properly on the A signal.

Telephone Business Office

When you are ready to install telephones, call the business office of the local telephone company, tell them your telephone number and that you are replacing existing rental telephones or that you are adding additional telephones. Give them the FCC information. They will tell you where to take your rental telephones. Check your next bill to see that your bill is reduced by the rental of the telephones you returned.

Party Lines and Coin Operated Telephones

Installation of your own telephones can only be made on single party lines. You may not install your own telephones if you are on a party line or if you have coin-operated telephones.

Tone Dialing Lines

If the existing telephones have a rotary dial and you want to replace them with tone-dialing telephones, you must call the local telephone company and ask them to connect you for tone-dialing service. There will be an extra charge each month. If you already have push-button tone-dialing telephones and install the electronic telephones that have push buttons but still do pulse dialing, everything will work properly. Pulse dialing works on tone-dialing circuits, but tone dialing does not work on pulse-dialing circuits.

Maintenance

The local telephone company will service any of their telephones and the telephone line to your building, but will not service telephones you have purchased. When a telephone service person is called to check a problem on your telephone, if they find trouble in the line to your building they should not charge you. If they find that the trouble is caused by telephones that you have installed, they will charge you a service charge. For this reason, you should plan your installation so that a central disconnect of all installed telephones is possible. In this way, if there is trouble in your telephone line after the installed telephones are disconnected, the telephone company will service the line. Presently this is done without charge, but conditions may change. Of course, if there is no trouble on the line but the trouble is in your installed telephones, you will have to return your telephones to where they were purchased and possibly pay a repair charge.

NOTES OF CAUTION

1. If you have a pacemaker don't install a telephone yourself. Call the telephone business office; they have a handicapped service that will help you. If you were to receive a shock from the lines due to ringing voltage or a lightning strike, it might affect your pacemaker.
2. This is the notice that appears in the instructions for installing a wall telephone modular jack distributed by Western Electric[2]: "CAUTION: Your telephone may have varying amounts of electricity in the wires and screws. Therefore, to avoid the possibility of electrical shock, follow the instructions below:

   If you have a telephone at a location other than the one you are removing, take the handset off the hook. (This will keep the phone from ringing and reduce the possibility of you contacting electricity. While you're doing this, ignore messages coming from the handset which will ask you to hang up the phone.) If you have only one telephone, take the handset of that telephone off the hook."
3. In your installation work, don't install your telephone while a thunderstorm is in process. Wear gloves wherever possible so you don't come in contact with bare wires. Handle bare wires with tools that have insulated handles. This is especially true for a screwdriver and long-nose pliers which are the tools used most often. Handle one wire at a time. Avoid contacting ground metal such as water pipes, radiators, heating ducts, etc. while in contact with telephone cable wires.

[2]Western Electric Instruction Bulletin CIB-2821-J2 JWR ©Western Electric Company.

4. Stay clear of the electrical power distribution lines running throughout your building when you do your installation. Do not mistake the electrical cable for telephone cable. The electrical cable is much larger and quite rigid. It requires quite a bit of effort to bend. Telephone cable is about the size of a straw used for drinking; it is very flexible and requires little effort to bend or twist. Electrical cable usually has two large insulated wires and a bare ground wire. Telephone cable has at least two wires, in most cases four wires, and in business and apartment installations may have upward of fifty wires.
5. Don't install telephones where a person using the telephone would be in a shower or standing in water or in a swimming pool. Water is a conductor of electricity and forms a connection if lightning were to strike or a short develop in a system.
6. Ordinarily when a telephone is not ringing, the electricity on the telephone line is such that, if a person touches the bare wires of the telephone line, it would cause nothing, or at best only a small tingle in a healthy person. It is when ringing voltage is on the line that more than a tingle occurs. Removing the handset to place the telephone off hook signals the central office that the line is busy.so that ringing voltage will not be applied to the line.

TOOLS

No extensive amount of tools or special equipment is required. However, before beginning, here is a list of tools that are needed:

TOOLS REQUIRED

(Most are required when adding telephones, fewer when just replacing telephones.)

A. -long nose pliers for handling wire leads
B. -diagonal cutting pliers for cutting wires
C. -medium weight screwdriver with at least a 1/4" tip and 5" shaft
D. -medium weight screwdriver with a 1/8" tip and 8" shaft
E. -tape for measurements
F. -hammer
G. -electric hand drill for up to 1/4" drills
H. -set of drill bits
I. -knife for stripping insulation
J. -wood spade bit 1/2"
K. -masonry bit 1/4" or 3/8" 18" long (called Bell Hanger bit)
L. -coat hanger

USEFUL TOOLS BUT NOT REQUIRED

A. - wire stripper, crimping tool for stripping insulation
B. - fish tape for in-wall installations
C. - keyhole saw for outlet holes
D. - lineman's pliers for pulling wires
E. - utility light
F. - square
G. - plumb bob for in-wall installations

CHANGING TO YOUR OWN TELEPHONES (MODULAR SYSTEM)

You have an apartment and you have assessed your present telephone system and sketched a floor plan as shown in *Figure 3-1*. The floor plan is just a rough sketch outlining the rooms and showing the position of the telephones. The apartment has been prewired with modular outlets and modular telephones are installed. There is a wall telephone in the breakfast room and an extension telephone in the bedroom. You want to install your own telephones at both locations. The wall telephone has a wall telephone modular jack plate already in place and the extension telephone has an outlet box with a modular jack faceplate *(Figure 2-6b)*. The outlet box is where the telephone line enters the apartment.

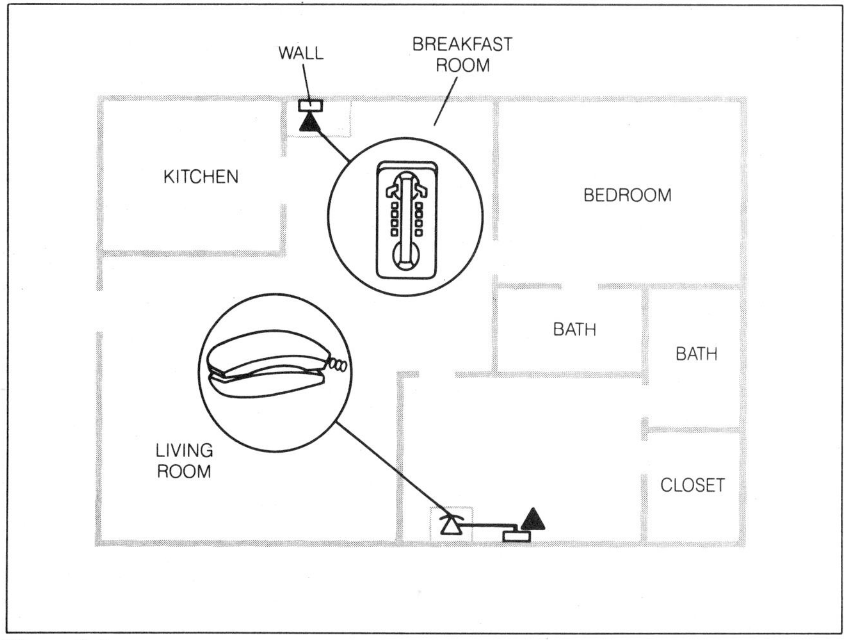

Figure 3-1.
Apartment Floor Plan (Modular System)

REPLACING EXTENSIONS IN MODULAR SYSTEM (RME)

Installing your own extension telephone into an existing modular system is very simple. The telephone comes with a cord that has a modular plug on the end. One end plugs in the telephone set and the other end plugs in the modular jack on the outlet box. Follow these steps:

Step	*Instruction*
1.	Unplug the modular plug of the rental telephone from the outlet jack by squeezing together the small bottom lever and the body of the plastic modular plug and pulling out on the cord as shown in *Figure 3-2a.*
2.	Unpack your purchased telephone. Plug one end of the modular cord in the telephone set.
3.	Plug the other end of the cord in the modular jack of the outlet as shown in *Figure 3-2b.* To make the connection, you are plugging your new telephone's modular plug in the same jack from which you unplugged the rental telephone.
4.	Lift the extension handset and you should have a dial tone.
5.	Other style extension telephones may have a modular cord that connects the handset to a base unit and another cord that connects the base unit to the telephone line. In such a case, you should be connecting the modular plug of the cord from the base unit to the modular jack of the outlet.

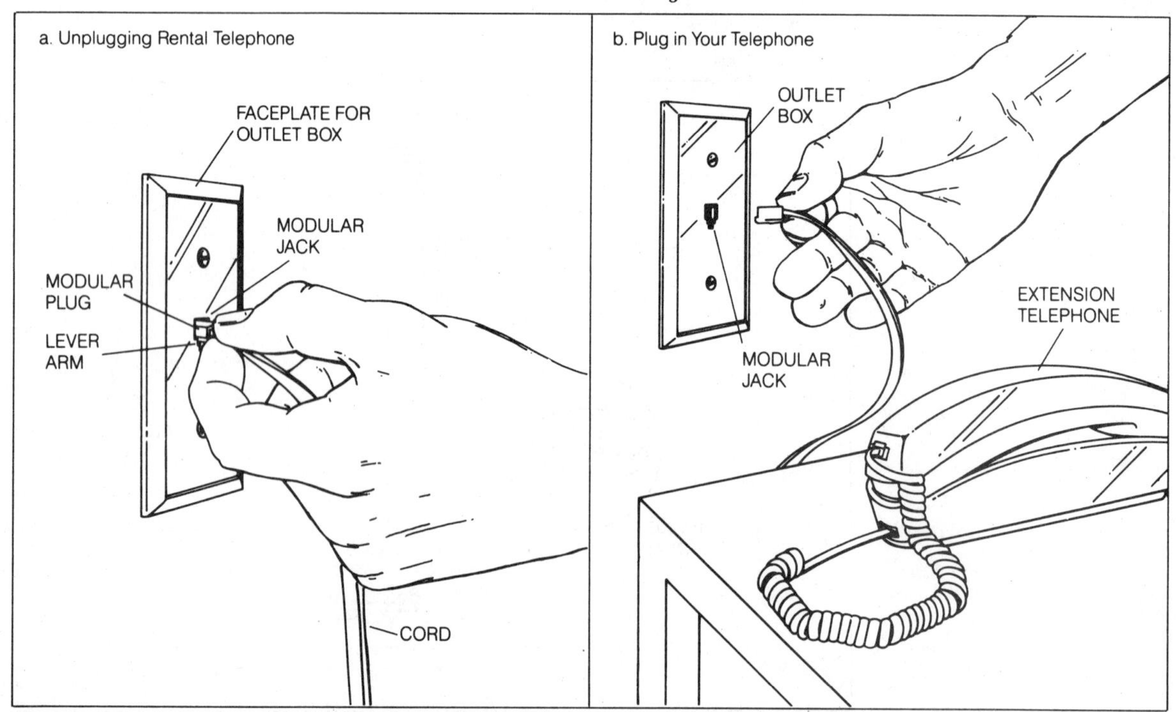

Figure 3-2.
Installing Modular Extension Telephone

REPLACING DESK OR TABLE TELEPHONES IN MODULAR SYSTEM (RMD)

Installing your own desk or table telephone (see *Figure 1-10*) when the existing system is modular is quite simple. The telephone that is used may have various ways that the cords attach. There may be a cord from the handset to the base unit that is connected permanently; or one end may be permanent and the other end have a modular connector; or both ends may have modular connectors. If the telephone has two cords, one connects the handset to the base unit and the other connects the base unit to the outlet jack. If the telephone has only one cord, then the one cord connects the handset to the outlet jack. The cord to connect to the telephone line will always have a modular plug on the end. In the apartment example there is no desk telephone, but if there were one, here is how it would be installed:

Step	*Instruction*
1.	Unplug the modular plug of the rental telephone from the telephone company from the modular jack in the outlet in the wall (or it may be on the baseboard) by squeezing together the small bottom lever and the body of the plastic modular plug and pulling on the cord as shown in *Figure 3-2a*
2.	Plug the end of the cord from the base unit that has a modular plug on it in the modular jack in the outlet mounted in the wall. This is the same as *Figure 3-2b* except the telephone is a desk or table model. (You are plugging your telephone's modular plug in the same jack from which you unplugged the rental telephone.)
3.	Lift the desk or table model telephone and you should have a dial tone.

Bringing Outlet Closer to Desk or Table

It may be that you do not wish the cord of the desk or table telephone to be loose back to the modular connector. In that case, a modular extension can be run along the baseboard and terminated in a modular jack mounted to the baseboard. The modular plug from the desk or table unit is then plugged in the baseboard modular jack. This installation is shown in *Figure 3-3*. Instructions 4 through 13 apply.

4.	Purchase a surface-mounted modular jack and extension cable of proper length with modular plug on one end and spade lugs on the other. If proper length does not fit standard cable, cut the cable to the correct length so that bare wire connections can be made in place of spade lugs. *Figure 3-4* supplies some hints on stripping and connecting bare wires.
5.	Plug modular plug of extension cable in the jack of outlet in the wall.

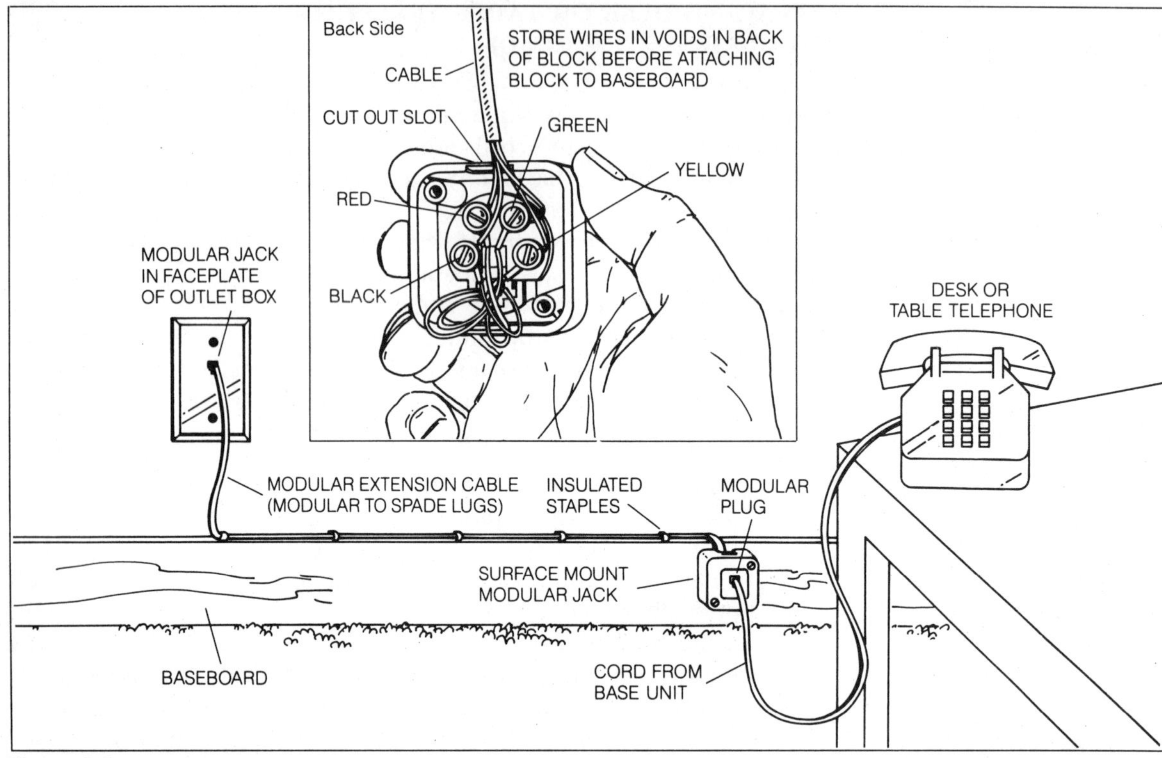

Figure 3-3.
Installing Baseboard Modular Connector

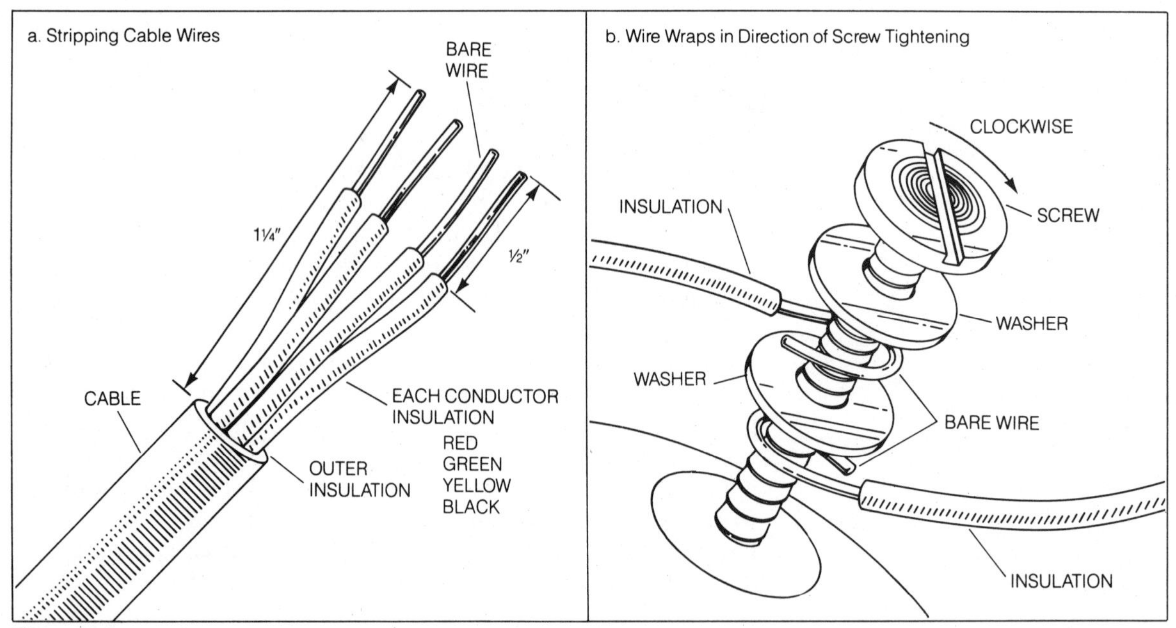

Figure 3-4.
Stripping and Wrapping Wire for Connection

6. Run extension cable down to baseboard and across top of baseboard to the position desired for surface-mounted modular jack.
7. On the back edge of the top of the surface-mounted modular jack, cut out a slot for the cable to run through. A sharp pocket knife works well for this.
8. Connect the ends of the extension cable, either spade lugs or bare wires,to the screws on the backside of the surface-mounted jack as shown in *Figure 3-3* red to red, green to green, yellow to yellow, black to black.
9. Staple the extension cable in place at the top of the baseboard with insulated staples, leaving enough slack to properly position surface-mounted jack.
10. Unplug the plug from the modular wall outlet.
11. Position surface-mounted jack against baseboard with extension cable out the top through the cut-out slot, and screw in place.
12. Plug in the modular plug for the desk or table telephone.
13. Replug the modular plug in the faceplate jack.
14. Lift handset of desk telephone. You should have a dial tone.

REPLACING WALL TELEPHONES IN MODULAR SYSTEM (RMW)

Installing your own wall telephone (see *Figure 1-10*) into an existing modular system is quite simple. The modular wall telephone comes equipped with a modular plug that plugs in a jack on a plate mounted on the wall. A pictorial sketch is shown in *Figure 3-6*.

Standard Wall Telephone

Step *Instruction*

1. To remove the rental telephone grasp the telephone as shown in *Figure 3-5* and push upward until the telephone slips from the mounting pins shown in *Figure 3-6*. Pull the telephone away from the wall plate. This disconnects the modular plug. (Some telephones have a slide clip or lever that must be released before the telephone can be slipped from the wall plate.)
2. To mount your own telephone, position the modular plug on the back of the wall telephone so that it will engage the jack in the wall plate as shown in *Figure 3-6*. Push the telephone flat against the wall with the plug in the jack. Raise the telephone until the mounting slots shown in *Figure 3-6* engage the mounting pins on the wall plate.
3. Using the same type grip on the telephone as shown in *Figure 3-5* push down on the telephone until it is firmly in place. (Snap the slide clip or lever in place to lock the telephone for those telephones that have them.)
4. Life the handset and you should have a dial tone.

More Contemporary Style

You may be interested in installing some of the more contemporary styled telephones rather than a standard type. Any of the ones with a modular connector on the back with the mounting slots would mount the same as described in steps 1 through 4. However, some of the other types require a little different technique. For example, the telephone shown in *Figure 3-7a* can be either a desk telephone or the same cradle can be used to mount the telephone to the wall with the modular wall plate that is in place. Such an installation, shown in *Figure 3-7b* and *3-7c,* can be made by following steps similar to 5 through 8.

Step	*Instruction*
5.	Remove the contemporary telephone from its box and wrapping. Hold it upright and look at the back. As shown in *Figure 3-7b,* there are two cords that connect to the back of the telephone with modular plugs. One connects the handset to the base, and the other is a line cord that would be plugged into a modular outlet if the telephone is used as a desk telephone.
6.	As shown in *Figure 3-7b,* the line cord that would be used if the telephone is used as a desk model is removed by pulling the cord out of the groove in the base. The cord runs in the groove so the telephone will set flat when used as a desk model. When the cord is out of the groove, unplug its modular plug from the modular jack in the mounting cradle.
7.	Obtain a 5¼″ modular line cord and plug the modular plug on one end into the modular jack from which the desk line cord was removed. As shown in *Figure 3-7c,* insert the modular plug on the

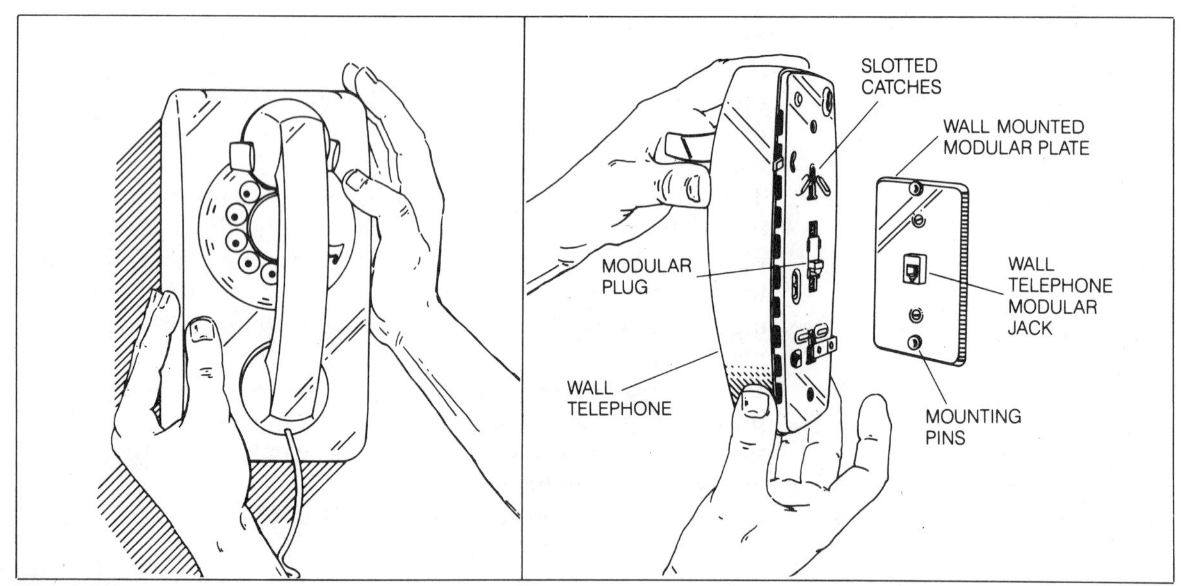

Figure 3-5.
Removing Rental Unit

Figure 3-6.
Modular Wall Telephone

other end into the modular jack on the wall plate that is in place. Store the excess cord in the rectangular well provided in the base, align the mounting slots in the base with the pins on the wall plate, press the telephone flat against the wall plate, and push down on the base until it is firmly in place. Hang the handset in the cradle base. Remove the handset; you should have a dial tone. If not, check to make sure plugs are in jacks properly.

8. The telephone should be mounted snuggly to the wall. If it is not, remove the base and screw in the mounting pins a turn or two (or

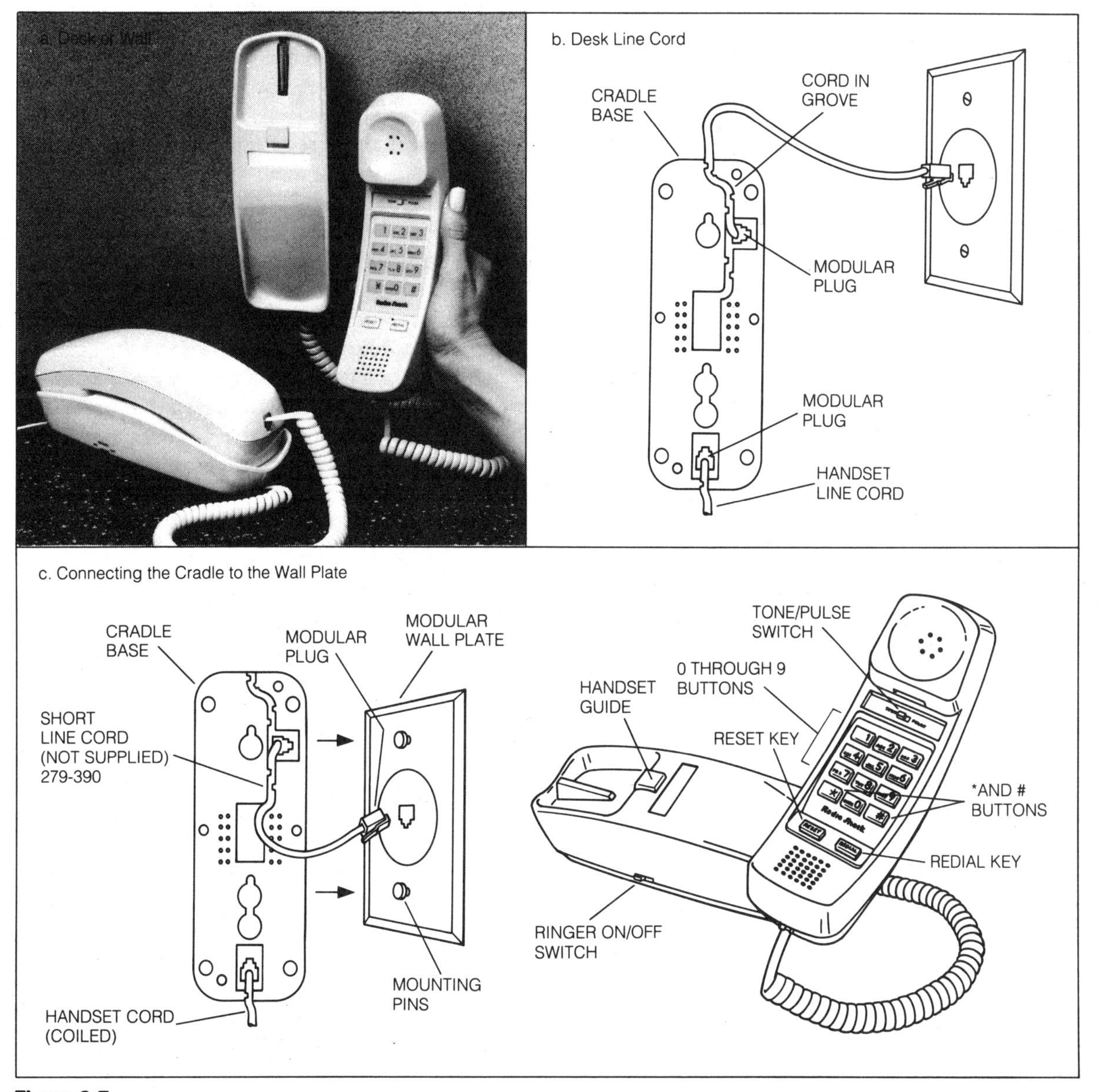

Figure 3-7.
Wall-Mounted Contemporary Telephone
Source: Radio Shack

perhaps you need to screw them out if the cradle is not engaging correctly). Remember to rotate the handset guide as shown in *Figure 3-8b* to hold the handset properly.

NOTE: When choosing a contemporary style telephone that can be used as a wall telephone, check the base carefully that is used for wall mounting. Not all bases will fit the mounting studs on the wall telephone modular plate.

No Wall Plate Mounting

9. Many of the contemporary telephones, like the one shown in *Figure 3-8*, can be mounted on the wall without a modular wall plate. Locate a wall stud and determine where the telephone should be located when mounted on the two screws provided. As shown in *Figure 3-8a*, locate the screws apart vertically 3 15/16". Drill small starting holes and insert the screws into the stud until the underside of the head is 3/16" from the wall.

10. As shown in *Figure 3-8a*, plug the modular plug on one end of the line cord provided into the back of the telephone and run the line cord in the groove in the base of the telephone so the telephone is flush with the wall when it is mounted on the screws.

11. Plug the handset cord into its modular jack. If there is a grove for the cord, place the cord in it. Hang the telephone on the screws and plug the other end of the line cord into a convenient modular outlet. Lift the handset off hook. You should have a dial tone. As shown in *Figure 3-8b*, pull up and rotate the handset guide to prevent the handset from falling off the cradle base now that the base if used as a wall telephone.

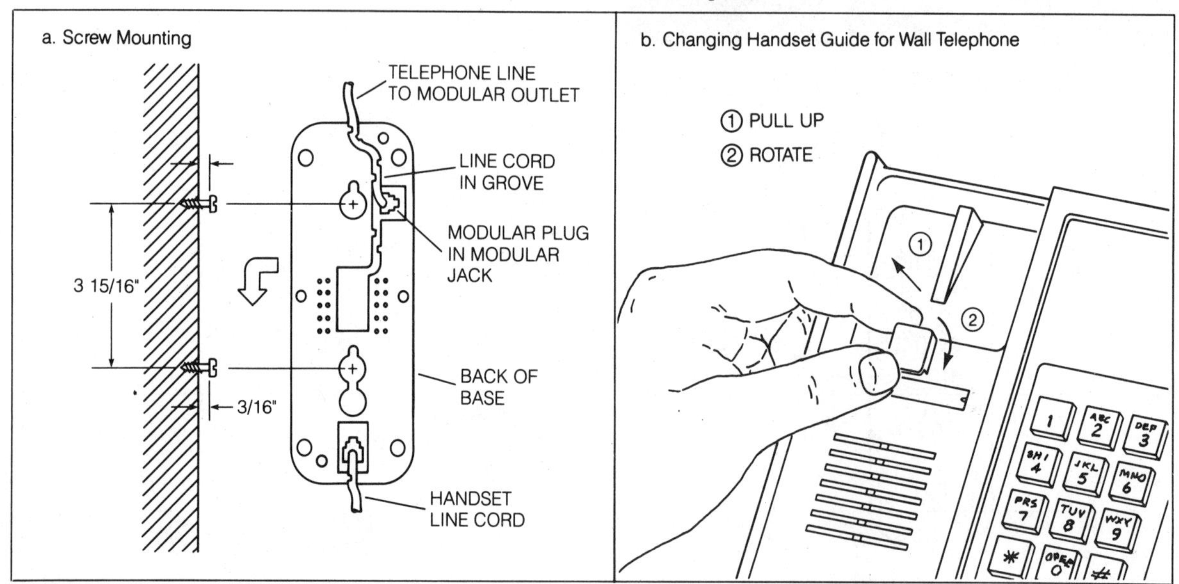

Figure 3-8.
Wall Telephone Mounted on Screws

A longer extension line cord may be desired or required when using an extension line cord for installing baseboard outlets *(Figure 3-3)*. If so, the extension line cords can be lengthened by using a line coupler that receives a modular plug from each side *(Figure 3-9)*. To install the coupler, plug the modular plug from the end of one of the extension line cords in one side of the coupler. The coupler may be attached to the baseboard with double sided adhesive tape. Plug one end of the second extension line cord in the other side of the coupler. The two extension line cords now are coupled together and can be used as one. If telephones are to be added to this system, refer to Chapter 5 and 6.

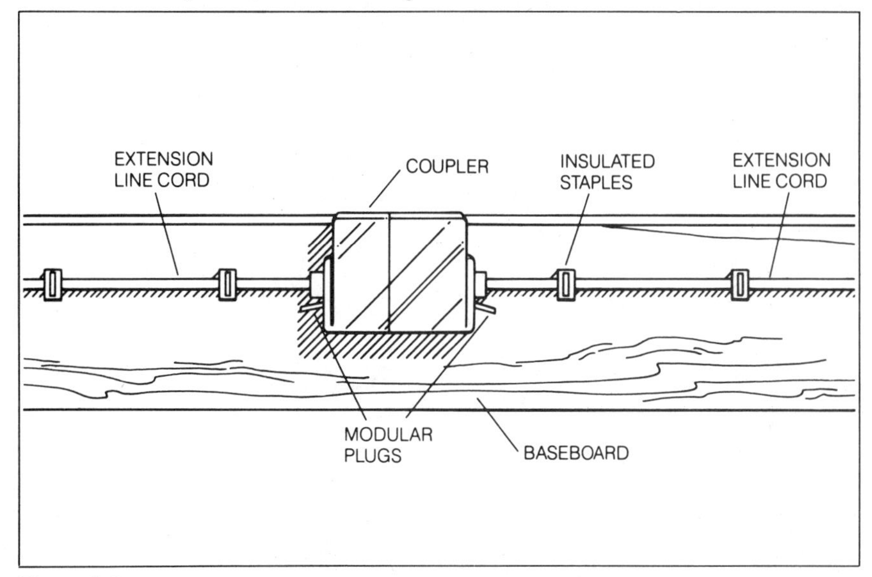

Figure 3-9.
Lengthening Extension Line Cords with Coupler

AREA CODE 502
L
K 5
J
M N 6
O
P 7 R S
8
TUV
9
WXY
0
OPERATOR

4 REPLACING OLD STYLE TELEPHONES

Chapter Content *Page*

Assessing System 2
Replacing Extension Telephone 3
Converting a 42A Block
Spade Lug Jack 5
Snap-on Modular Jack 6
Converting 4-Prong Plug 7
Cord Adapter 7
Surface-Mount Jack 7
Distribution of Modular Telephones 8
Replacing Desk or Table Telephone 8
Using Cord Adapter 9
Changing Outlet Jack 9
Replacing Wall Telephone 12
Removing the Telephone 12
Installing Wall Plate 15
Making the Connection 15
Finishing the Job 18

Older homes and apartments will have present telephones interconnected with old style connectors. Replacing existing telephones with your own telephones requires that the connections be converted to modular connections because the purchased telephones will have modular connectors. Universal jacks, adapter plugs and jacks, and snap-on connectors can be used to make the conversions.

After reading this chapter you should:

A. Know how to convert 42A blocks to modular jacks.
B. Know how to disconnect directly-wired telephones and connect modular telephones.
C. Know how to convert 4-prong jacks and plugs to modular.
D. Know how to convert directly-wired wall telephones to a wall telephone with modular connectors.

Let's look at how old style systems can be converted.

ASSESSING AN OLD STYLE SYSTEM

You live in a home and have assessed your present telephone system and sketched a floor plan as shown in *Figure 4-1*. The rough sketch shows the position of the wall telephone at the end of the cabinet in the kitchen, the old style 4-prong jack just above the counter in the kitchen and the 42A block mounted on the baseboard in the hall. The dotted line on the plan shows that the telephone line runs from a pole in the front yard overhead to a protector mounted on the soffit *(Figure 2-3a)*, up into the attic where it is connected to a 42A block being used as a terminal strip for a junction point. From the 42A block, one line goes across the attic and down into the wall to the kitchen outlet box which has a 4-prong jack faceplate as shown in *Figure 4-7*. The other line runs from the 42A block, across the attic, through the ceiling in a closet, down the closet wall, through the closet wall facing the hall, and into the hall just above the baseboard. It connects to a 42A block mounted on the baseboard in the hall. The interconnection is diagrammed in *Figure 4-2*. The extension telephone in the hall connects to the 42A block with spade lug connections, while the line from the attic has insulation stripped from its wires and the bare wires wrapped around the screw terminals to make connection (See also *Figure 3-4b)*.

A cable runs from the kitchen counter outlet up through the wall into the overhead kitchen cabinet, under a shelf to the end of the cabinet, through the end of the cabinet to an old style rotary-dial wall telephone (*Figure 3-5*).

REPLACING AN OLD STYLE EXTENSION TELEPHONE (ROE)

Installing your own extension telephone like the one shown in *Figure 3-2* is quite simple. The telephone comes with a modular plug on the end of a cord ready to connect to a modular jack. The only thing that needs to be done is to convert the 42A block in the hall to a modular jack. (Actually something else must be done if you install the specific telephone into a system that has had only rotary-dial telephones. If you want to operate with tone dialing, you must call your local telephone company and ask to be connected for tone-dialing service).

Converting a 42A Block to a Modular Jack

Step	*Instructions*
1.	(Recall that if you are disconnecting existing telephones, remove the handsets so the telephones are off hook. This will prevent the telephone from ringing and keep you from getting a mild shock.) To replace the rental extension telephone, the telephone cord must be disconnected from the 42A block. The easiest way to do this is to snip the wires of the telephone cord at the 42A block.

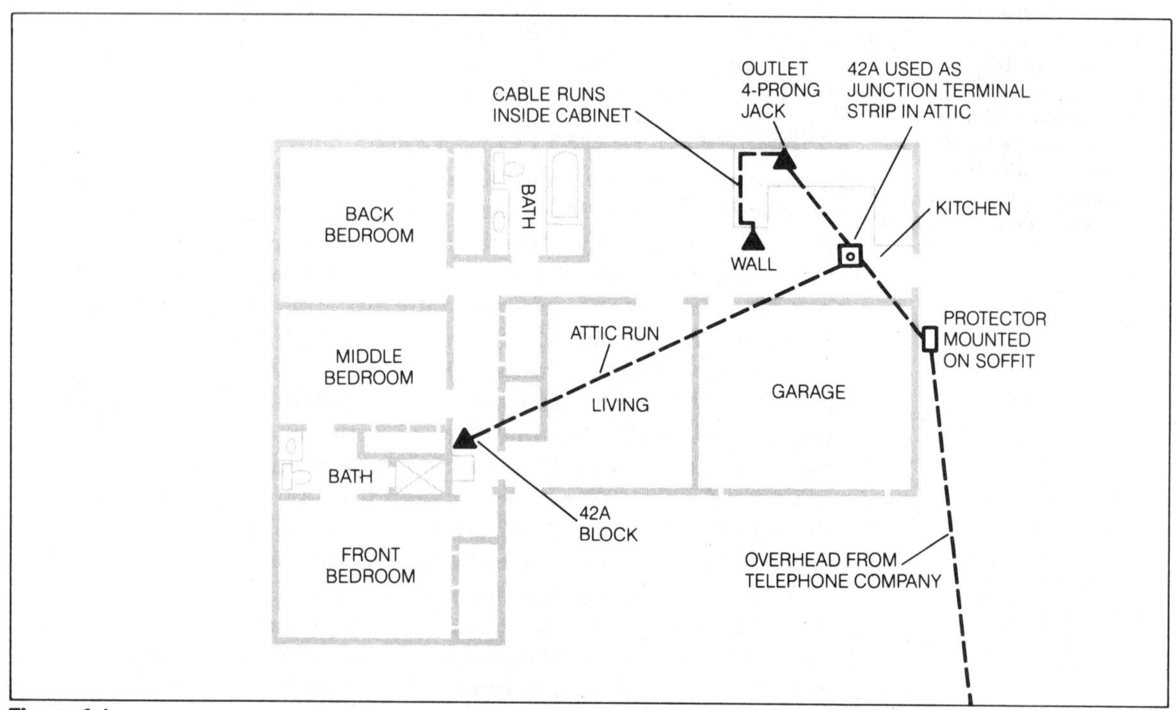

Figure 4-1.
Floor Plan of House with Old Style Telephone Connections

CAUTION: FOR EXTENSION (OR DESK) TELEPHONES THAT HAVE A LIGHT IN THEM.

If the extension (or desk) telephone that you are disconnecting has a light in it, there is an ac voltage being supplied from a transformer through the yellow and black wires to the light bulb in the telephone set. The transformer is one that is plugged into a standard ac outlet nearby and wires run from the transformer to the 42A block to connect to the yellow and black leads of the cable to the telephone set. This transformer is identified in your telephone operator's manual. Know what this transformer looks like so that you can unplug it and remove the ac voltage from the 42A block before you snip any leads.

Snipping the Leads

With an insulated screwdriver, take off the cover of the 42A block. With insulated diagonal cutters or insulated scissors, snip the telephone cord at the four terminals as shown in *Figure 4-3*. The telephone cord wires are the ones with the spade lugs. Cut the wires at the end of the spade lug as they come from under the screw holding them in place. Don't cut any other wires just the telephone cord wires.

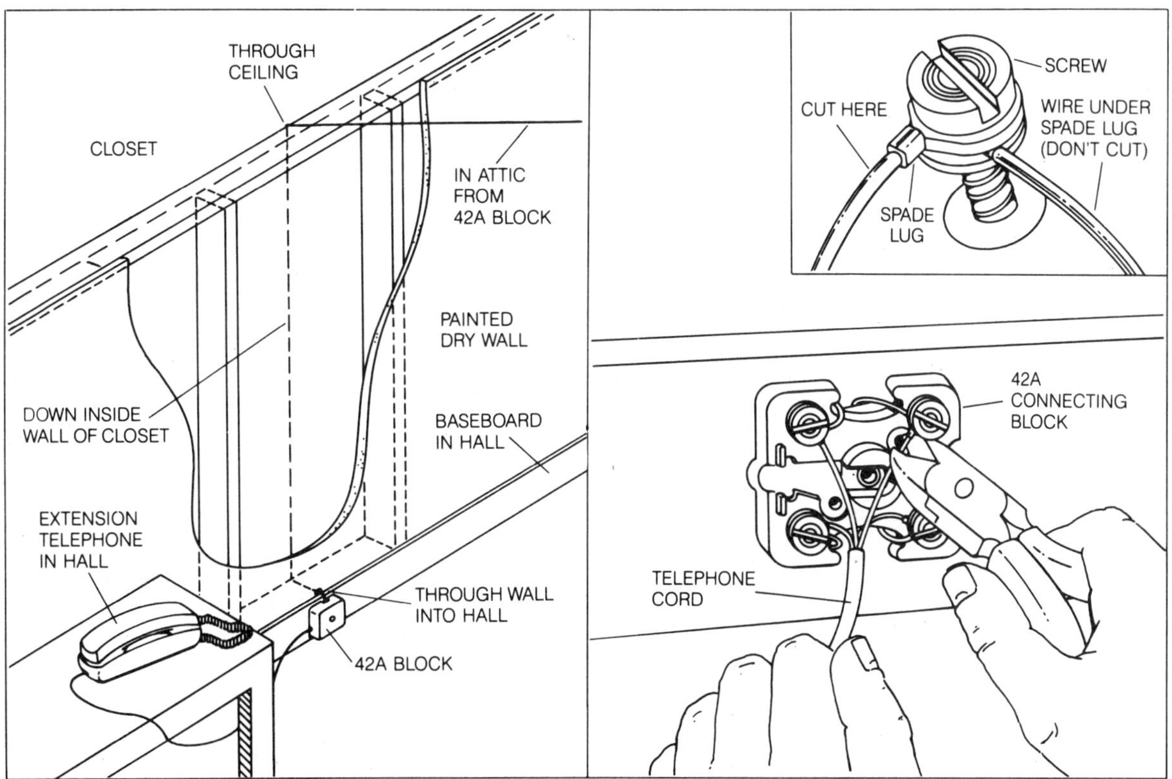

Figure 4-2.
Line Run for Hall Telephone Extension

Figure 4-3.
Disconnecting Telephone Cord

Using a Modular Jack with Spade Lugs

2. An easy way to convert the 42A block to a modular jack is to use a "Quick-Connect" jack and install it as shown in *Figure 4-4*. Remove the cover from the 42A block and discard. The modular jack forms a new cover. Make a cutout in the side of the quick-connect modular jack to allow clearance for the incoming telephone cable to the 42A block when the modular jack is fastened over the 42A block.
3. Disconnect the rental telephone by loosening the screws on the 42A block terminals and slipping out the spade lugs that are connected to the wires of the rental telephone. If you have difficulty getting the lugs loose, just snip the wires at the lugs as previously described. Insert the spade lugs of the modular jack wires under each terminal, respectively, and tighten the terminal screw. Note that the four wires are color coded red, green, yellow and black. Match the colors of the wires at each terminal: red is connected to red, green to green, yellow to yellow and black to black. Your installation should look similar to *Figure 4-4a.*
4. The quick-connect modular jack can be positioned over the 42A block in any of four positions. As you face the 42A block, it is best to position the jack so that the modular connection is made from the right, left or bottom. Making the connection from the top exposes the jack to dirt, dust, and debris that could cause poor connection.

   Decide on the most convenient position for the jack, position it over the 42A block, align the screw in the top of the jack with the screw hole in the 42A block and fasten the modular jack to the 42A block with the screw provided *(Figure 4-4b).*

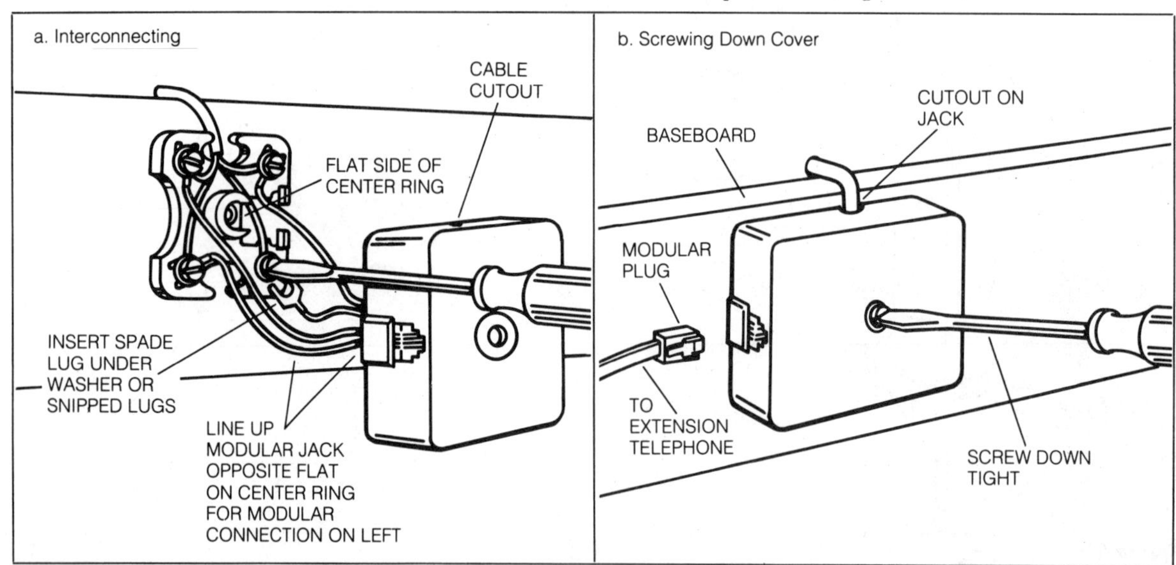

Figure 4-4.
Installing Quick-Connect Jack

5. Connect the modular plug on the end of the cord of your purchased telephone in the modular jack. You should have a dial tone when you lift the handset. If there is no dial tone, unfasten the modular jack and carefully examine each connection. Make sure the color code connections are correct. Make sure there are no exposed wires touching each other. If you snipped the wires, one of the lugs from the snipped wires can easily move when you are tightening the screw terminals and short out to a neighboring wire.

Using a Snap-On Modular Jack

The 42A block can be converted to a modular jack very easily using such a snap-on modular jack.

6. With an insulated screwdriver remove the cover of the 42A block. Discard it. Disconnect the rental telephone by snipping the wires as in step 1.
7. The modular jack forms a new cover for the 42A. Inside the new cover are color coded snap-on connectors. Snap them on the screw terminals as shown in *Figure 4-5a*. Each snap is color coded to match the wires on the 42A — red to red, green to green, yellow to yellow and black to black.
8. Screw on the new cover. Plug in the modular plug of the telephone cord as shown in *Figure 4-5b*.
9. Lift the telephone off hook. There should be a dial tone.

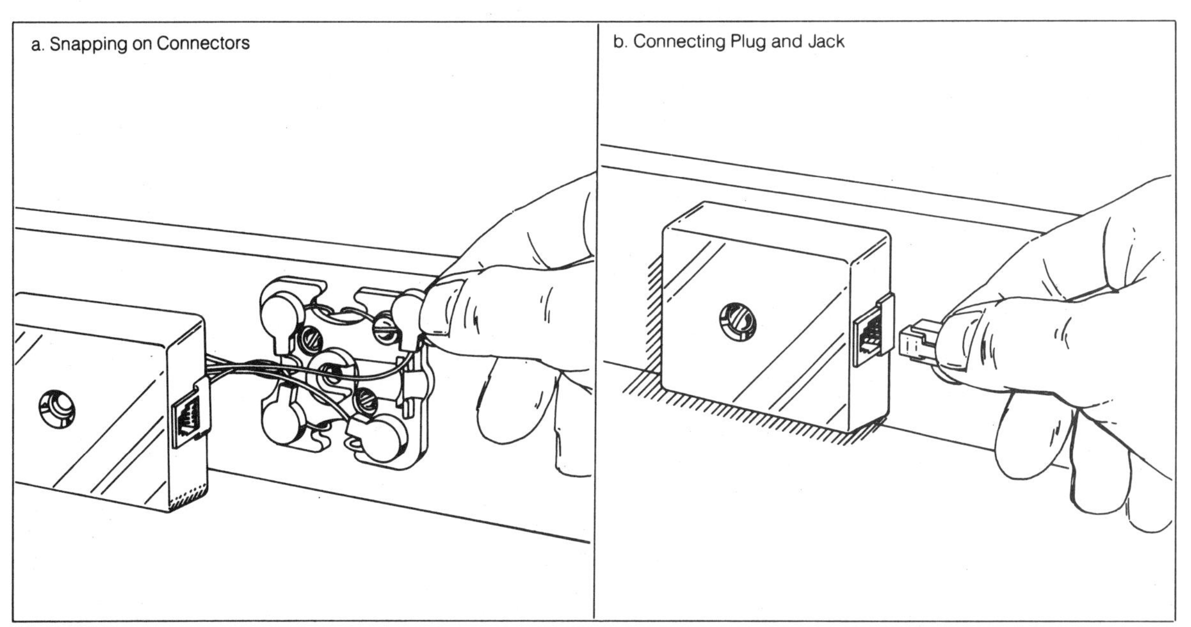

Figure 4-5.
Installing Modular Jack with Snap-On Connectors

Converting a 4-Prong Plug to a Modular Jack

Using Cord Adapter

If the extension telephone that you are replacing is an old-style telephone with a 4-prong connector, then the outlet on the baseboard will be a 4-prong outlet as shown in *Figure 4-6*. There is an easy way to convert the 4-prong outlet to a modular jack without disturbing any connections. As shown in *Figure 4-6a*, the conversion is done by using a cord adapter that converts the 4-prong outlet to a modular jack. The conversion is done in three simple steps.

Steps	*Instructions*
1.	Disconnect the old-style rental extension telephone.
2.	Plug in a 4-prong cord adapter which has a modular jack within it.
3.	Insert the modular plug from your purchased extension telephone into the jack, lift the handset and you should have a dial tone.

Using a Surface-Mount Modular Jack

If the extension telephone old-style outlet is only an extension outlet, then the 4-prong outlet can be converted to a surface-mount modular jack (*Figure 2-10c*). This is done by removing the extension telephone cable connections from the 4-prong old-style outlet and rewiring them to the surface-mount jack as described previously for *Figures 3-3* and *3-4*. Your purchased extension telephone is plugged into the surface-mount modular jack to make your extension telephone active.

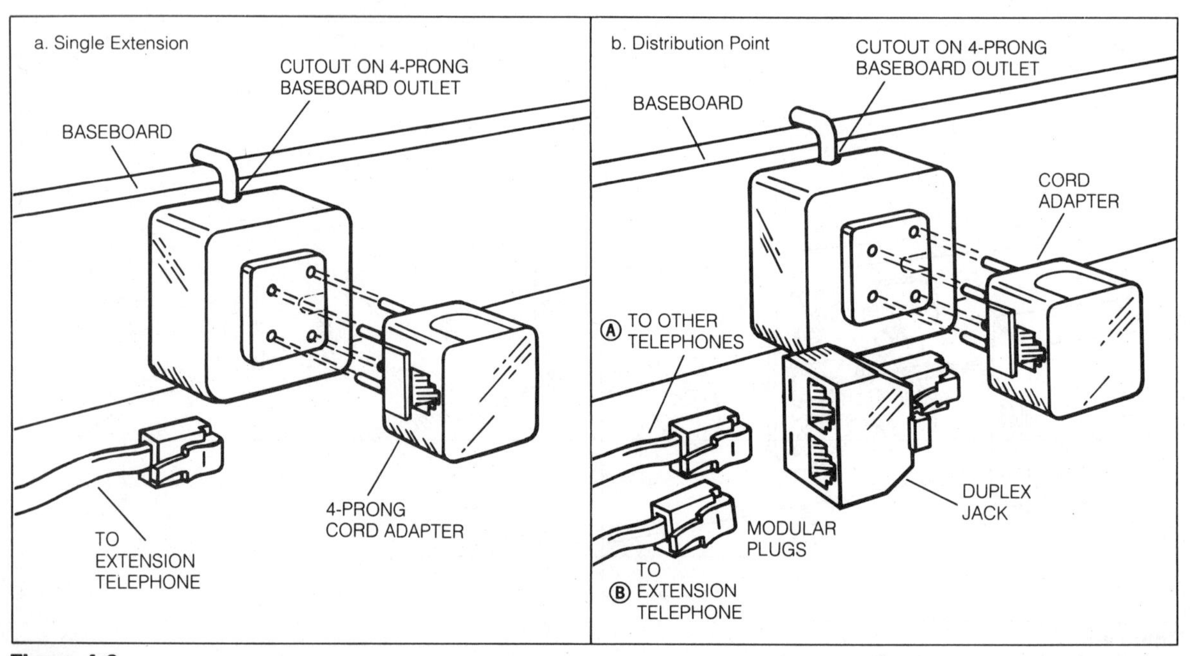

Figure 4-6.
Converting 4-prong outlet to Modular

Distribution to Modular Telephones

If the old-style outlet is the first initial entry into a house or apartment bringing the telephone company's line inside, then this outlet can be used as a distribution point to other modular telephones. As shown in *Figure 4-6b*, a duplex jack can be plugged into the cord adapter that has been plugged into the 4-prong old-style outlet. One modular jack of the duplex jack is used for the extension telephone and the other jack is used for a modular outlet to all other telephones in the home.

Convenient Connection for Troubleshooting

This is a particularly convenient connection because an extension telephone is connected to one jack and all other telephones to the other jack. If there is trouble on the line, such an arrangement becomes easy to troubleshoot. All the other telephones are connected to the modular plug A; therefore, with A disconnected, all other telephones are removed from the line. If there is still trouble on the line, the extension telephone can be removed by unplugging B. Unless there is trouble with the duplex or adapter jacks (and these could be replaced with new ones), any trouble still on the line must be due to the incoming line and is the responsibility of the local telephone company.

REPLACING AN OLD STYLE DESK OR TABLE TELEPHONE (ROD)

In the house plan layout there is an outlet with a 4-prong jack above the counter in the kitchen. Before the wall telephone was installed at the end of the kitchen cabinet, this outlet with its 4-prong jack was used for a desk rotary-dial telephone shown schematically in *Figure 4-7*. Normally, when an outlet like this is used to provide the connection for a desk or table telephone, the 4-prong jack is mounted in an outlet box that is up from the floor about 12 to 18 inches rather than being on a counter. This is shown in *Figure 4-7* as the dotted line connection to the desk telephone.

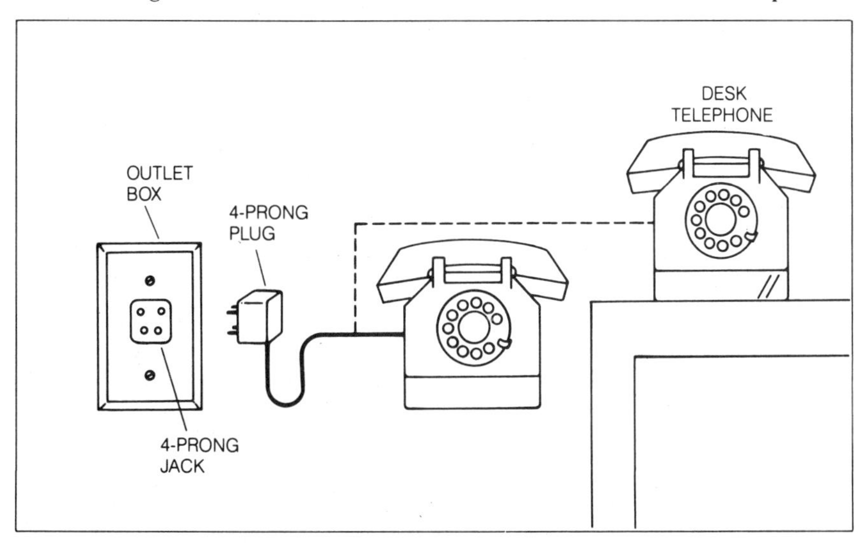

Figure 4-7.
4-Prong Jack Connection for Desk, Counter or Table Telephones

Converting 4-Prong Jack in Outlet Box to Modular

There are a number of ways to convert a 4-prong jack mounted in an outlet box to a modular connector.

Using a Cord Adapter (Modular Plug Converter)

The most direct way is to use a cord adapter that converts the 4-prong jack to a modular jack by plugging in the cord adapter. This was shown in *Figure 4-6*. Plug the cord adapter into the 4-prong jack in the outlet box and you have available a modular jack into which a modular desk telephone can be connected.

Changing the Outlet Box Jack and Faceplate to Modular

A second way is to rewire the outlet box jack and substitute a modular jack and new faceplate. The 4-prong jack in the outlet box is wired as shown in *Figure 4-8*. Follow these steps to convert it to modular: (This example uses spade lugs or bare wires.)

Step	*Instruction*
1.	(Remove handset so one telephone is off hook.) With an insulated screwdriver, remove the two screws that hold the faceplate to the 4-prong jack bracket.
2.	Remove the two screws that hold the bracket to the outlet box. Pull the bracket forward so the screw terminals are available on the back of the jack body *(Figure 4-8)* and disconnect the cable wires from the terminals. The terminals should contain red wires connected, yellow wires connected, green wires connected and black wires connected. (In some cases, only one wire may be connected to each terminal.) Do this carefully. Don't touch any bare wires. Don't break any wires.
3.	Discard the bracket and 4-prong jack.
4.	Obtain a modular jack assembly that mounts in the outlet box and has a faceplate that fits flush to the wall. One of these modular jacks and its faceplate is shown in *Figure 4-10*.

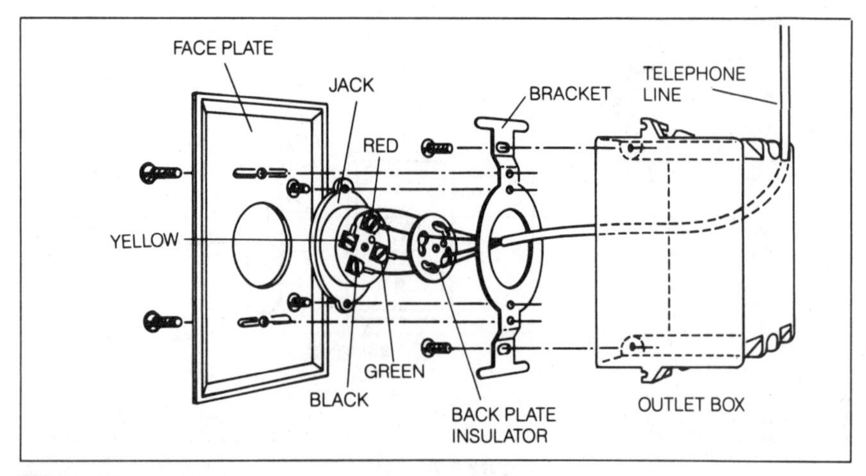

Figure 4-8.
4-Prong Jack and Faceplate Mounting

5. The assembly of the modular jack to its bracket shown in *Figure 4-10* is very similar to the 4-prong jack shown in *Figure 4-8*. Hold the bracket with the jack body so that the screw terminals are facing you as shown in *Figure 4-9* and connect the cable wires from the outlet box to the screw terminals. If there are spade lugs, put the spade lugs under the screw washers. If there are bare wires, wrap the wires around the screw terminals and tighten the screws *(Figure 3-4b)*. Connect red to red, green to green, yellow to yellow and black to black.
6. Push the cable back into the outlet box and align the bracket holding the modular jack with the mounting holes in the outlet box. Using the long screws supplied mount the bracket to the outlet box. The modular jack is mounted so that the release lever slot is down and the connecting rails are up.
7. Complete the assembly by attaching the faceplate to the bracket, fitting it over the modular jack body and adjusting it to fit flush to the wall and square before the two faceplate screws are tightened. The completed assembly looks like *Figure 2-10a*.
8. The modular jack is installed and ready to receive a modular plug from a newly purchased telephone.

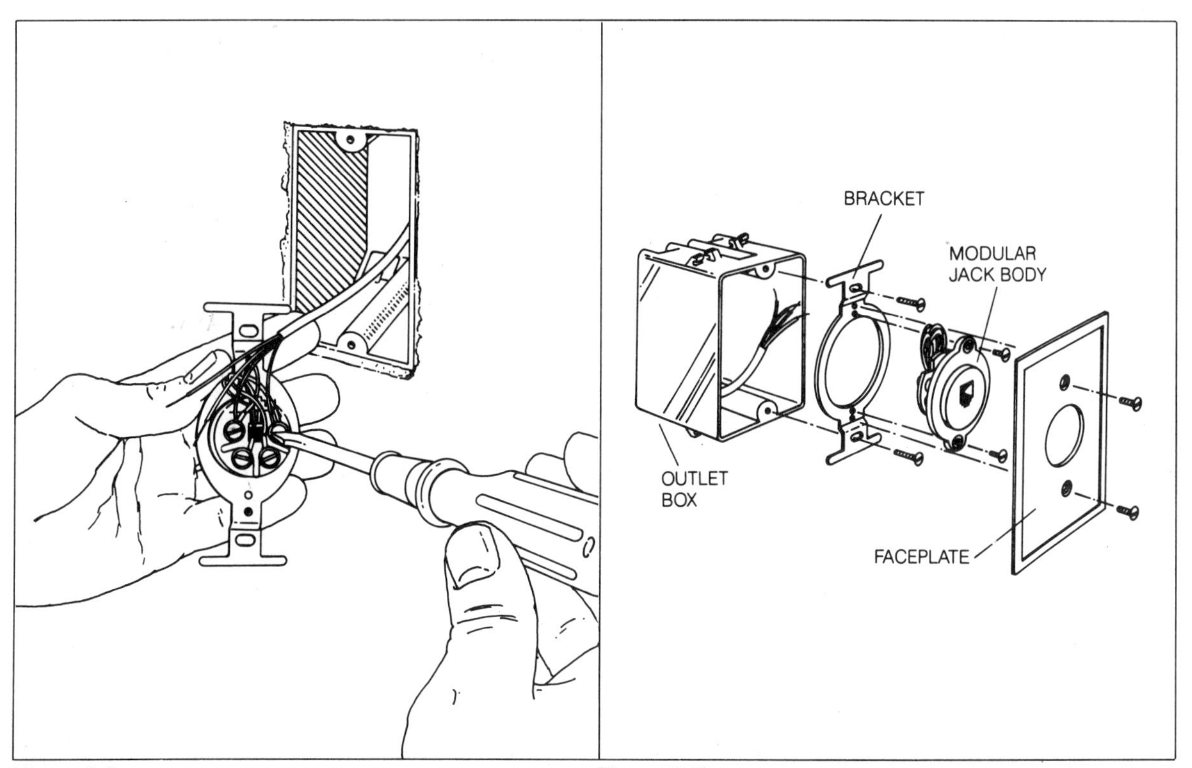

Figure 4-9.
Connecting Wires to Modular Jack

Figure 4-10.
Rectangular Modular Outlet

Connecting a Desk Telephone With a 4-Prong Plug to Modular

Suppose you had changed to a modular jack and faceplate on an outlet box and had turned your old style rental telephones back to the telephone company. Now you have an opportunity to buy a second-hand 4-prong telephone set (or maybe a friend gives you one), then a modular four-prong adapter shown in *Figure 4-11* could be used to connect the 4-prong plug of the telephone set to the modular jack. Plug together the male plug from the telephone set and the female plug adapter. Then insert the modular plug in the modular jack of the outlet and the telephone is installed.

Other Old Style Connections For Desk Telephones

Desk telephone cords may go directly into an outlet box through a faceplate that has a hole in the center of it. Such a connection was described in *Figure 2-5b*. A 42A block was inserted into an outlet box. The telephone set cord connects to the 42A block with spade lug connections as shown in *Figure 4-12*. *Note in particular the connection of the yellow wire from the telephone set to the green wire at the 42A block.* This was the way the

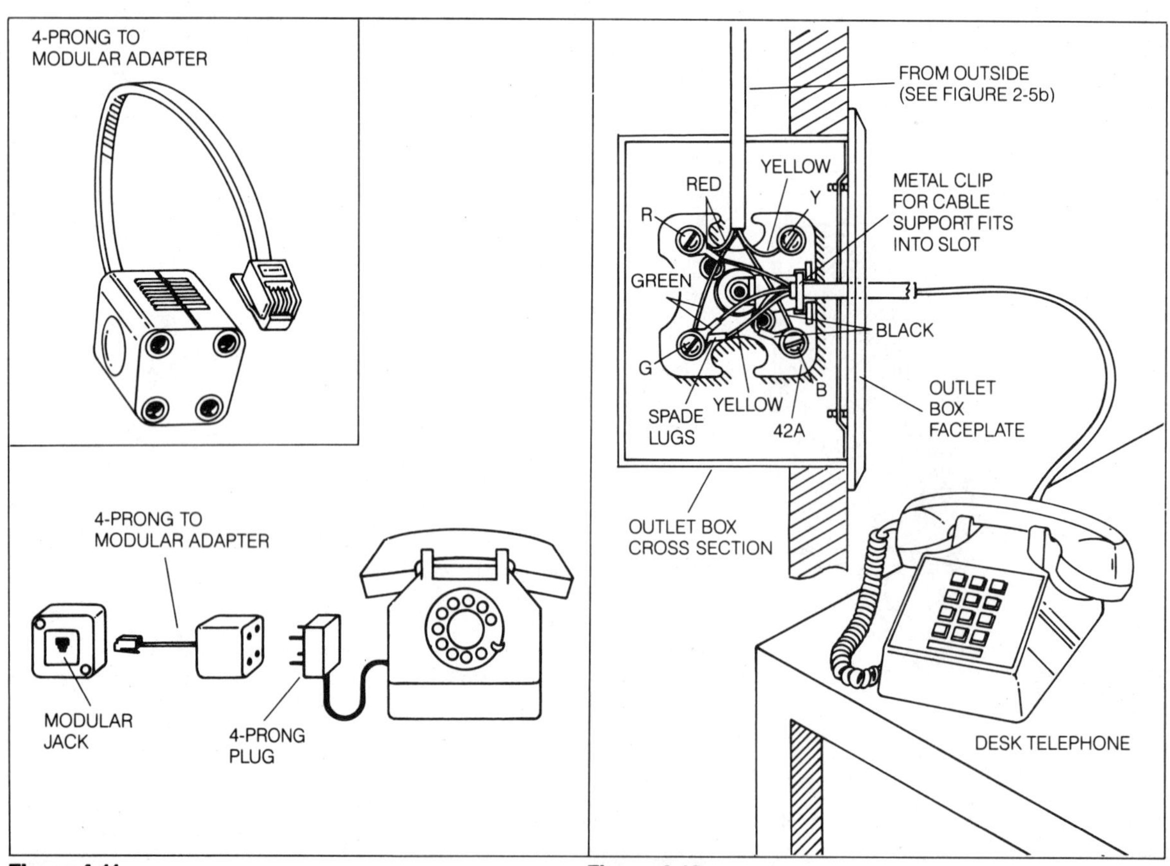

Figure 4-11.
4-Prong to Modular Adapter

Figure 4-12.
42A Connections for Desk Telephones

telephone installer used to control whether the installed telephone would ring or not. If the yellow wire is connected to the green wire the telephone will ring. If the yellow wire is connected to the yellow wire from the outside, not to the green, then the telephone will not ring. Some modular connected telephones have this option available at the telephone set not the connecting jack.

The cord from the telephone set also may go directly to a 42A block mounted on a baseboard. This type connection was shown in *Figure 4-2*.

In either case, the connections to the 42A block are shown in *Figure 4-12*.

To convert the 42A block connections of this type to modular, follow these steps:

Step	*Instruction*
1.	Remove the 42A cover with an insulated screwdriver.
2.	Either snip the leads with an insulated diagonal cutter as in step ROE-1 or loosen the screws and pull out the spade lugs of the telephone cord. Note the metal clip shown in *Figure 4-12* that slips into a slot on the 42A block to provide cable support so the wires are not pulled loose. Pull the snipped four conductors or the wires with the spade lugs free and return the telephone to your telephone company.
3.	Tighten the screw terminals on the 42A.
4.	Use any of the previous procedures for converting the 42A block to a modular jack outlined in the ROE discussion (steps 1-9).
5.	Your newly purchased desk or table telephone is installed by plugging the modular plug in the newly installed modular jack.

REPLACING OLD STYLE WALL TELEPHONES (ROW)

The telephone at the end of the cabinet in the house plan layout of *Figure 4-1* is a wall telephone in which the telephone cable runs into the case of the telephone and is connected directly to the speech network (See *Figure 4-14*). To change a telephone like this to a modular telephone follow these steps:

Removing the Telephone

Step	*Instruction*
1.	(The following instructions are for a common type of wall telephone. There are other types that have screws that hold the case in position.) Remove the case of the wall telephone by taking a screwdriver and pushing up on a spring clip at the bottom of the telephone case as shown in *Figure 4-13*. (The eraser end of pencil makes a good tool for this.) There is a u-shaped notch in the case and a portion of the spring clip is exposed. Pushing up on the spring clip and in on the bottom of the case allows the case to be released and pulled away from the the plate connected to the wall

as shown in *Figure 4-13*. Remove the handset from its hanger and unhook the case from a spring latch at the top of the back plate. Then lift the case over the hanger so it is free to be put aside.

2. Remove the telephone from the wall by unloosening two screws *(Figure 4-14)* that hold the back plate to the wall and slip the back plate off of the screws. The back plate has elongated slots for the two screws, one at the top and one at the bottom, so that the back plate will slip off when the screws are loosened. Behind the back plate there should be an outlet box. The telephone cable wired directly to the wall telephone comes from the outlet. There may be cases where there is no outlet box and the cable is just coming through a hole in the wall. In the floor plan that we are dealing with, the cable comes out of a hole in the cabinet over the counter.
3. Lift the back plate containing the main body of the telephone from the support of the screws and find the cable that makes the telephone connection. Trace the cable into the telephone and find where it connects to the telephone. It should be connected to the speech network as shown in *Figure 4-14*. The incoming red wire is connected to L2 on the speech network and the incoming green wire is connected to L1. The black and yellow wires are left unconnected. They are used as a spare pair in case problems develop in the red and green pair.

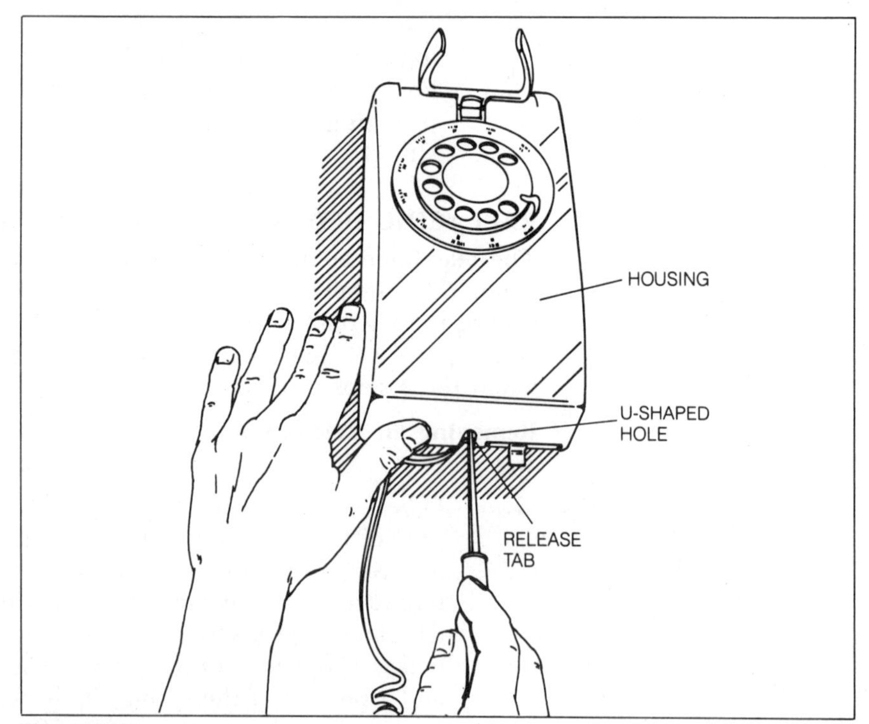

Figure 4-13.
Releasing Spring Clip

Put the telephone back on the wall by placing the back plate over the screws for support. This is to support the telephone while you disconnect the line.

4. (Recall that to prevent a mild shock in case the telephone should ring take the handset off the cradle. This takes the telephone off hook. Disregard the message from the central office to hang up the telephone, you'll have the line disconnected in a minute.) The easiest way to disconnect the line is to snip the wires with an insulated diagonal cutter. Make the cut right at the screw terminals as shown in *Figure 4-14*. Of course, you could loosen the screws and take the wires loose. Loosening the screws of the terminals and pulling the wires free is a simple task in itself and leaves conductors with bare wire so that no insulation stripping is necessary when connecting the wires at the next step. If the telephone is being used as a junction point for the take off of a pair of wires to another telephone, disconnect the additional pair of wires as well.
5. Remove the old style telephone, put on the case and return it to the local telephone company.

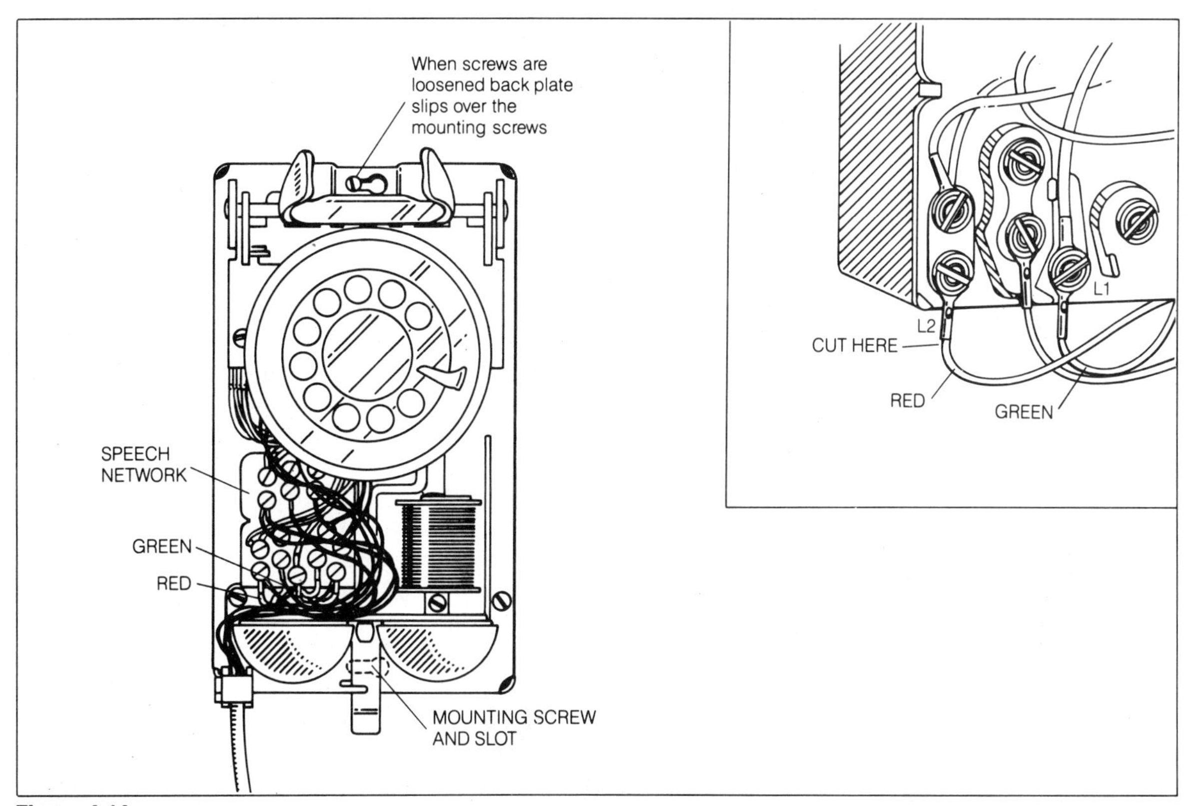

Figure 4-14.
Incoming Line Connected to Speech Network

Installing a Wall Telephone Modular Plate

6. You now have a cable with four conductors coming out of a hole in the cabinet at the end of the counter. For other wall telephone installations the cable comes out of an outlet box. In some cases the wall telephone is the junction point for an additional telephone; therefore, two four-conductor cables come out of the outlet box or from a hole in the wall.
7. Purchase a wall phone plate (wall telephone modular jack). It is the wall plate with a modular jack that is shown in *Figure 2-9* that accepts wall-mounted telephones with a modular plug on the back. Separate the faceplate from the back plate as shown in *Figure 4-15*. Put the faceplate aside. Insert the four conductor cable(s) through the opening provided in the back plate as shown in *Figure 4-15*.
8. Determine the position desired for the new wall telephone. Position the back plate with the modular jack facing away from the cabinet (or wall as the case may be) and the back plate indicating up (*Figure 4-15*). The back plate must be positioned vertical; otherwise, the wall telephone will not hang straight. Attach the back plate to the cabinet with the screws provided. If the back plate were being installed over an outlet box, the screw holes are matched to the outlet box spacings. Use screws that are normally used to attach electrical outlet plugs. If the back plate is being attached directly to a wall other than wood, use plastic anchors for the screws or molly bolts to hold the back plate in place.

Making the Connection

9. If sufficient outer insulation covering of the cable has not been removed, remove about 3″ of it as shown in *Figure 4-15*. This is the outer sheath, not the insulation for the individual wires. Included in the wall phone plate are plastic connector caps, one of which is shown in *Figure 4-16*. This plastic connector cap is used to connect the wires to the modular jack back plate. Two wires can be connected at once. The insulation need not be off the individual wires, the connections on the back plate are crimp connectors that bite into the wire and cut through the insulation.
10. Hold one of the connector caps with the long tab toward you and drape the wires over the connector as shown in *Figure 4-16*. Put the green wire in position 1 and the yellow wire in position 3. Repeat with the second connector cap putting red in position 2 and black in position 4. Fold the wires over the plastic connector from back to front as shown in *Figure 4-16*. Form the wire down around the connector on both back and front so the wire fits into the groves. Leave some excess wire and slack as shown in *Figure 4-16*. Of course, if the insulation is already removed from the individual wires because the wires were removed from the speech

network terminals by loosening the screws, the bare wires can be draped over the connector cap in the same way as the wires with the insulation.

11. **A** There are clear plastic flaps covering the back plate connectors. Fold these back so that the connector caps with the wires can be slipped over the back plate connector as shown in *Figure 4-16*. Do this with the first connector cap. Slip it onto the bottom of the back-plate connector so that it connects the same color coded leads. If you start with the green and yellow, repeat with the black and red connector cap. As shown in *Figure 4-16*, use a screwdriver as a lever to push the connector caps onto the back-plate connector. *The connector caps must be pushed all the way on. Make sure of this.* The connector cap should go up half-way on the back-plate connector.

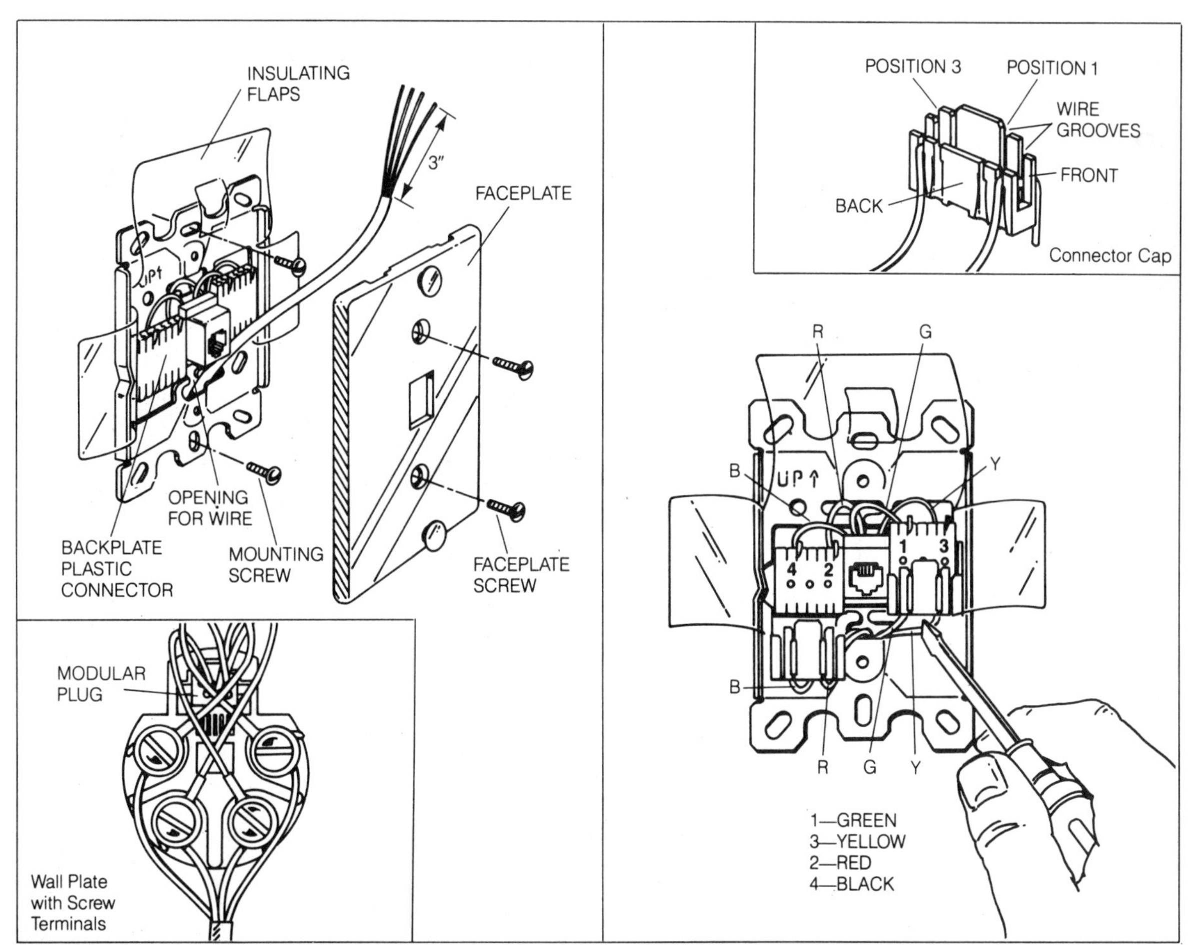

Figure 4-15.
Backplate and Faceplate

Figure 4-16.
Pushing on Plastic Connector

11. **B.** The connector caps are designed to give trouble free installation; however, if a connector cap is lost or if it is easier for you, the conductors can be pushed into the back plate connector slots with a screwdriver as shown in *Figure 4-17*. Use a screwdriver with a thin tip so that the connector will not crack. Make sure the conductor is pinched between the metal fingers in each slot. The insulation doesn't have to be stripped from the wire, but it may be a good idea to do so to make sure the bare wire is being pinched correctly. Some wall telephone modular jack plates include a small plastic tool to make the above insertion, or for making the insertions shown in *Figure 4-16*.

11. **C.** Other style wall plates may have screw terminals as shown in the insert in *Figure 4-15*. Strip the insulation from each conductor and connect the bare wires to the screw terminals the same as for *Figure 4-9*.

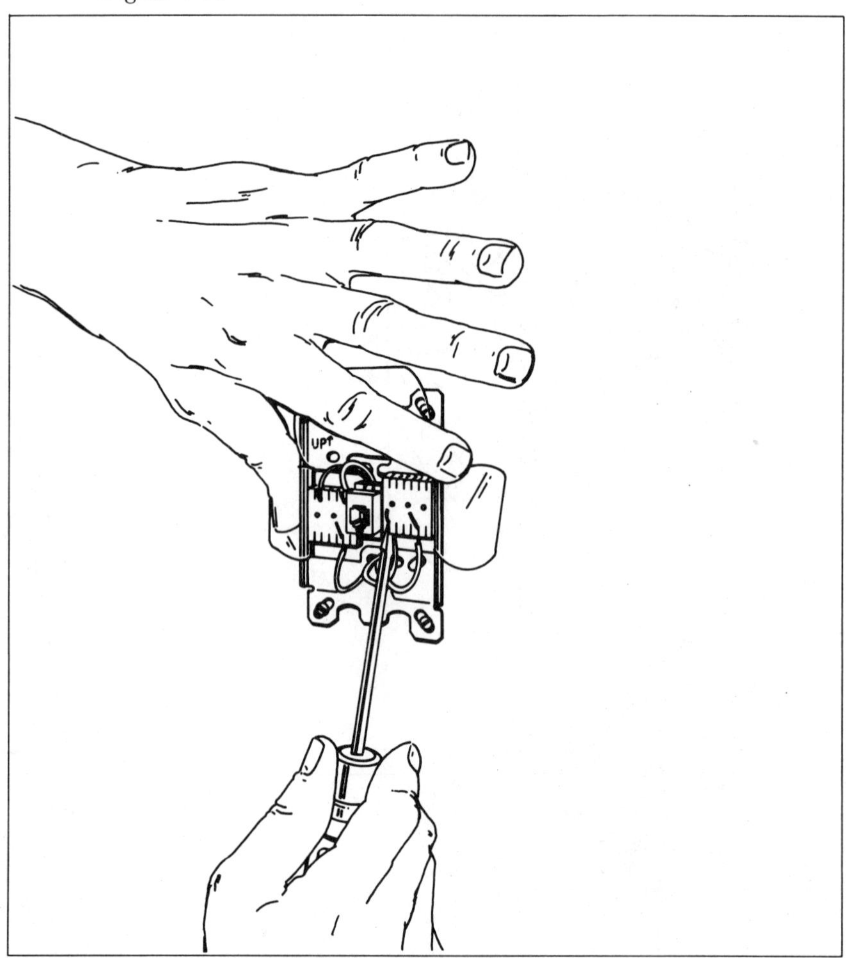

Figure 4-17.
Pushing Conductors into Connector with a Screwdriver

Finishing the Job

12. With insulated diagonals trim off excess wire from the connector caps. Push back any excess cable into the cabinet, outlet box or wall. Fold back the plastic sheets over the back plate connectors.
13. Join the face plate to the back plate and fasten with the screws provided as shown in *Figure 4-15*. Make sure that none of the wires are being pinched between the face plate and the back plate.
14. If the face plate is covered with plastic, peel off the plastic sheet.
15. Mount a modular telephone to the installed modular wall plate as described in steps 2 through 4 of the RMW instructions in Chapter 3.

For Exposed or Surface-Mounted Wiring

Figure 4-15 shows notches in the face plate. These notches on the top or bottom edge provide for mounting a modular wall plate when the wiring is not in the wall. The exposed wiring fits through the notches of the face plate and runs to the back plate to be connected with connector caps.

If Installed Tone-Dialing Telephone Won't Dial

One further item on tone-dialing telephones. If you know that the telephone company has made the connection for tone dialing service and your tone-dialing telephone will not dial but you can receive calls, reverse the connection of the tip and the ring. Change the red and green wires (if that is the color code of your system). Put the red wire where the green one is and the green wire where the red one is. The telephone should perform correctly.

Remember that many of the electronic telephones are pushbutton telephones but are not tone-dialing; they are still pulse dialing. The above problem is not corrected by a reverse connection on tip and ring for pulse dialing telephones.

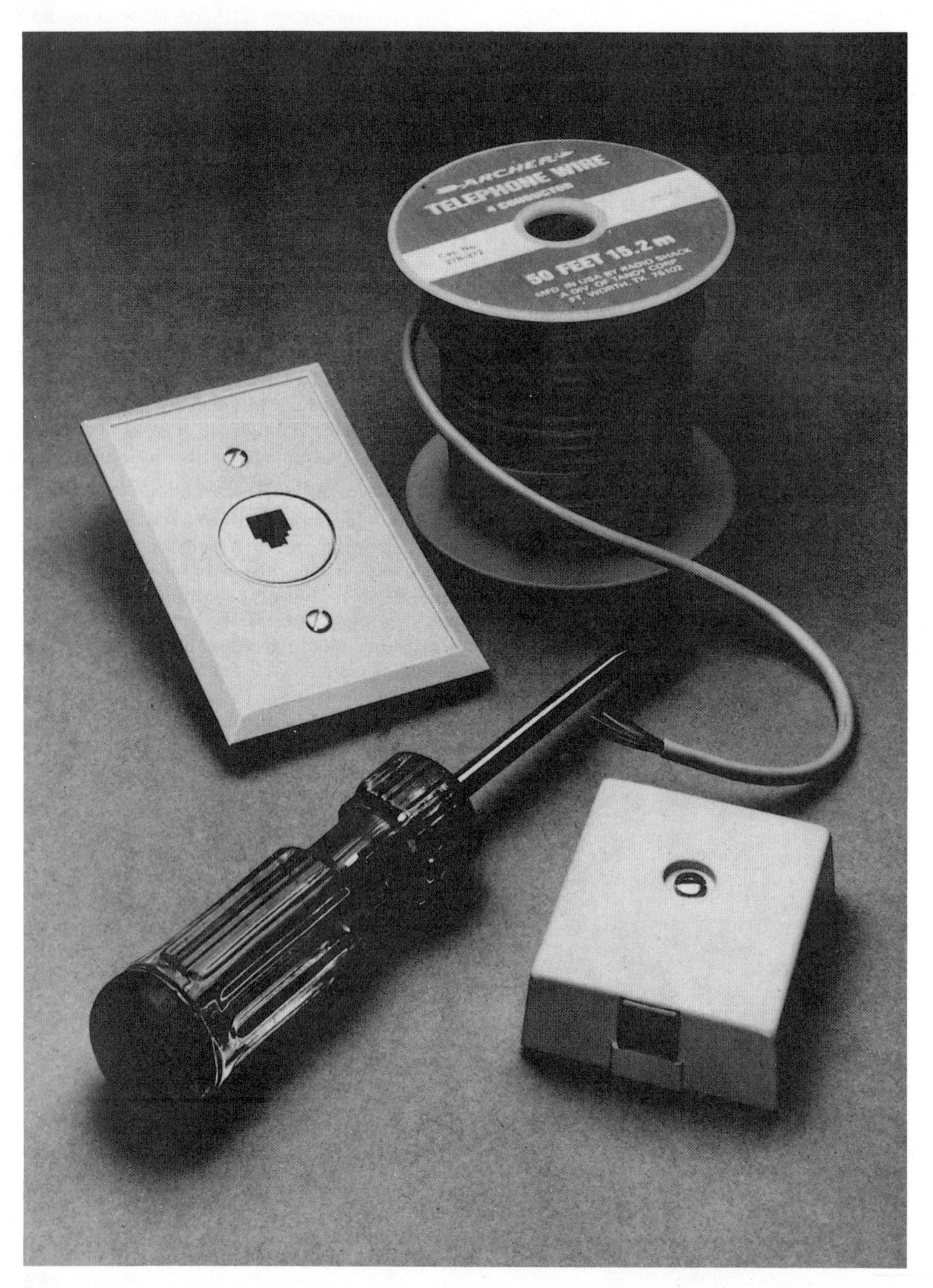
ARCHER
TELEPHONE WIRE
4 CONDUCTOR
50 FEET 15.2 m
MFD IN USA BY RADIO SHACK
A DIV OF TANDY CORP
FT WORTH, TX 76102

5-1

5 RUNNING INTERCONNECTING CABLES

Chapter Contents	*Page*
Common Cables	*2*
General Guidelines	*3*
Exposed Wiring	*3*
Concealed Wiring	*7*
Exterior Wiring	*14*
Trailers	*15*

Whether you live in an apartment, a house, or have a small business, there always seems to be a need to add a telephone. Some telephones can be added by buying couplers and extension cords; but for more permanent installations it is necessary to run telephone cable. This chapter shows how to install cable to make the telephone line connection to new or moved telephones.

After reading this chapter you should:

A. Be able to run exposed cable along baseboards, around door frames, inside closets, inside cabinets, etc.

B. Be able to run concealed cable inside walls by using attic or basement distribution.

C. Learn some tips about running cables in mobile homes.

COMMON CABLES

The most common telephone cable is a four conductor cable — a two pair cable. Two types are shown in *Figure 5-1*. Each type has four color-coded insulated solid-wire conductors inside an outer sheath of insulation. The modular cable has a flat rectangular cross section and the regular cable has a round cross section. The most common color coding is red, green, yellow and black for the four conductor colors. Other color codes are shown in *Table 5-1* for four-conductor (2 pair) and six-conductor (3 pair) cables.

Conductors					
4		6		6	
R*	T**	R	T	R	T
Red Yellow	Green Black	Red Yellow Blue	Green Black White	Blue Orange Green	White/Blue Band White/Orange Band White/Green Band
		Red Yellow Blue/White Band	Green Black White/Blue Band	Blue/White Band Orange/White Band Green/White Band	White/Blue Band White/Orange Band White/Green Band

Table 5-1.
Common Cable Color Codes

*R = RING
**T = TIP

SAFETY PRECAUTIONS

There are general safety precautions for cable installation. Important points from the National Electric Code are good guidelines. Turn to Chapter 9 and read the section on installation standards and codes before you begin.

GENERAL GUIDELINES

Two important points that have been mentioned previously need to be mentioned again before planning the layout of your added telephones. New telephones that you purchase and install will not be serviced by your local telephone company; therefore, plan your added telephone layout to have a central modular disconnect. This provides quick isolation of your added telephones when a trouble must be located in the telephone company line. A telephone wiring block that provides a junction for two cable connections of four conductors each (*Figure 5-2*) can be used. It connects to the system with a modular plug to provide an easy disconnect system.

EXPOSED WIRING — INTERIOR

The easiest cable to install for interconnections is cable that runs on the surface of walls, baseboards, around door frames, in cabinets and closets because the major runs are not concealed inside of walls. This type wiring is called interior exposed wiring.

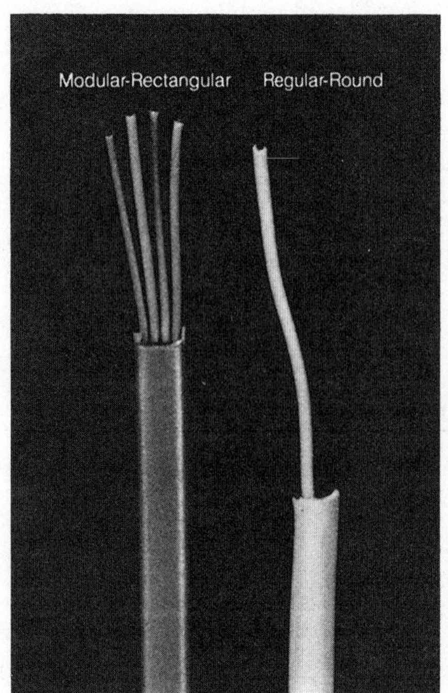

Figure 5-1.
Telephone Cables
Courtesy of Radio Shack

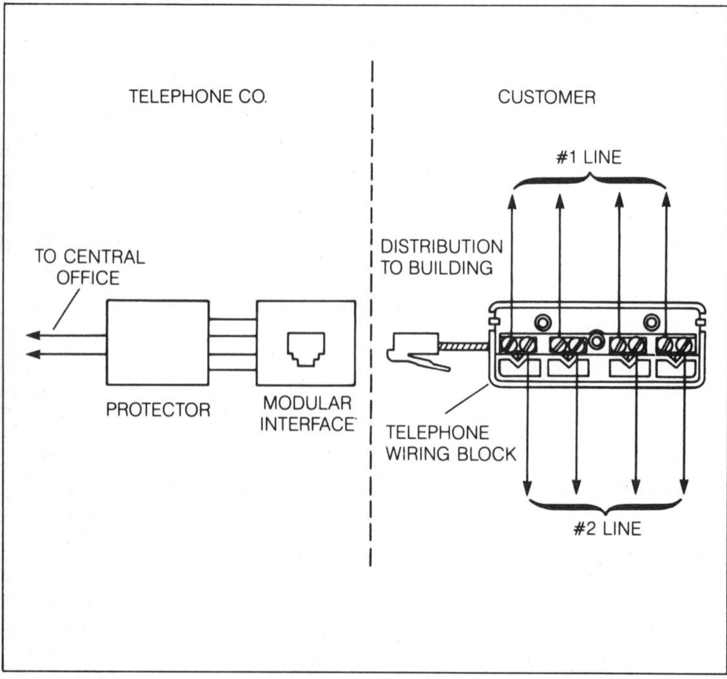

Figure 5-2.
Telephone Wiring Block

Baseboard and Door Frames

A very common way to run cable for an added telephone is to attach it to a baseboard or to a door frame. Take for example the arrangement in the bedroom of *Figure 5-3*. Point A is an outlet modular jack (*Figure 2-10a*) that was installed in the apartment when it was pre-wired. A new modular extension outlet is desired at point B. Cable is run from point A to point B along the baseboard and around the door frame. It terminates in a surface-mounted jack (*Figure 2-10c*) that is screwed to the baseboard. The cable is made neat and trim against the top of the baseboard, and is held in place with insulated staples that are anchored in the baseboard or door frame. The cable can be attached to screw terminals on the jack inside the outlet box at point A. A small notch can be filed in the outlet face plate to allow the cable to come out the bottom on the wall surface. The modular jack at point A is still available to receive a telephone connection.

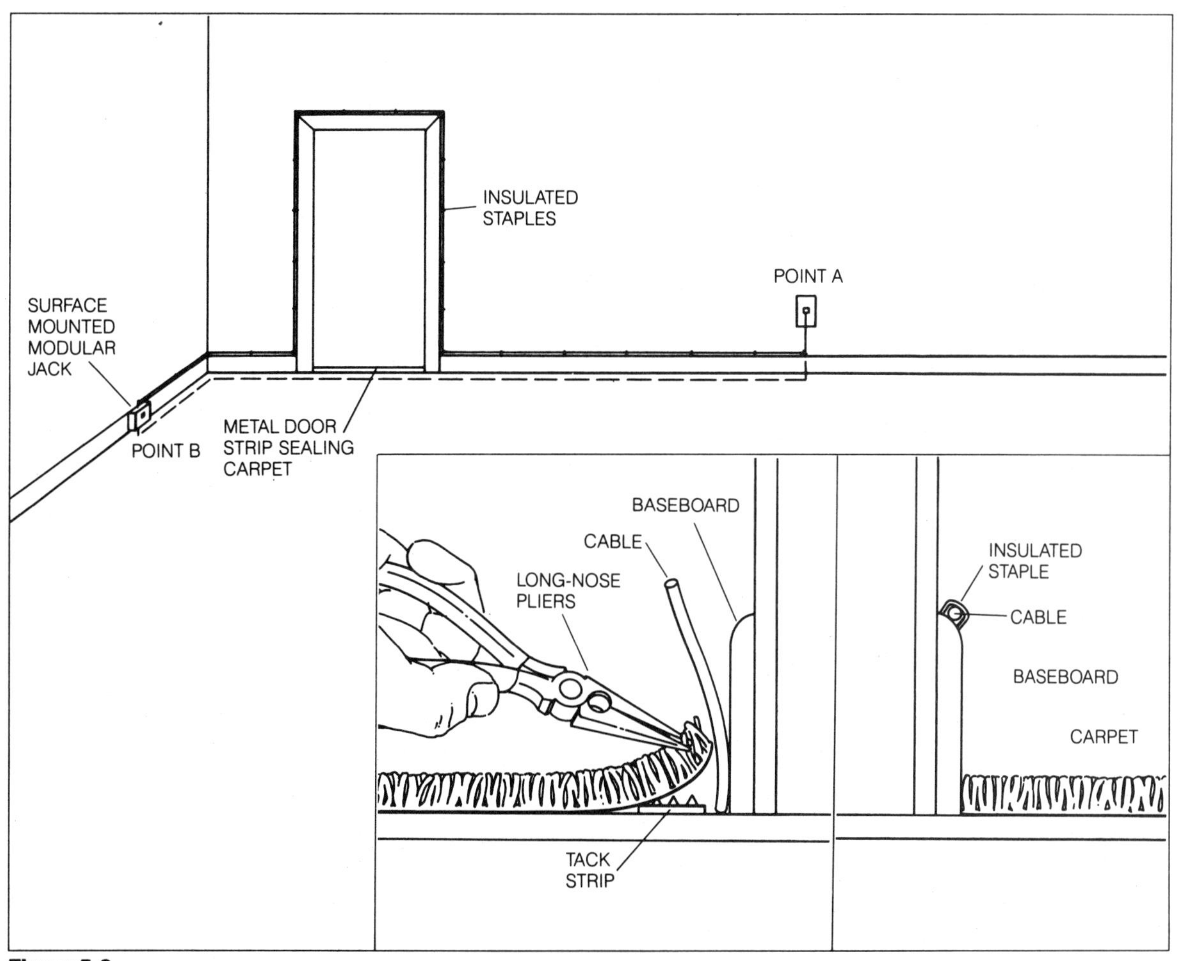

Figure 5-3.
Cable Runs Along Baseboard and Doorframe or Under Carpet

Under Carpet Runs

One of the neatest ways to run the cable without exposing it is shown in *Figure 5-3*. This is the same installation as on the baseboard and around the door frame except now the cable runs along the baseboard under the carpet. The carpet is pulled up with a long nose pliers as shown in *Figure 5-3* without pulling the carpet away from the tack strip. At the doorway, the cable runs under the metal strip that finishes the carpet in the doorway.

Cabinet Runs

Cable can also be run in cabinets in order to keep it from being exposed. In *Figure 5-4* there is a round modular outlet at point A. It is used for an extension telephone. A wall phone is desired at point B. A new cable is connected to the screw terminals of the round modular outlet, brought out the bottom through a filed notch, run along the top of the wainscoat trim, up the side of the splashboard, and over the top edge until it enters the cabinets attached to the wall.

Inside the cabinet, the cable is run in the front and top corners to keep it hidden. A lower corner of the end of the cabinet is selected to allow the cable to exit on top of the splashboard, run around the edge of the splashboard and the wainscoat until it comes to the door frame. At the door frame, it travels up until it is brought across horizontally to the modular wall plate for the wall-mounted telephone. As shown in *Figure 4-15*, there are notches in the faceplate of the wall telephone plate to clear surface-mounted wires.

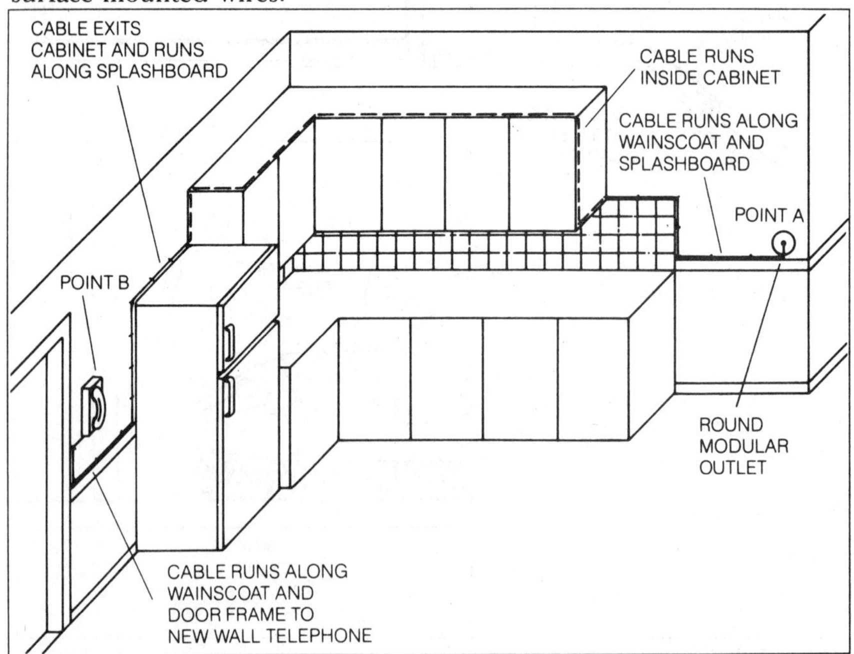

Figure 5-4.
Cable Runs Inside Cabinet

Closet Runs

Closets also are good places to run exposed surface cable to keep the cable hidden. Point A of *Figure 5-5a* is an outlet modular jack (*Figure 2-10a*). A round outlet modular jack is desired at point B. This outlet can be mounted directly in sheetrock or drywall by using plastic screw anchors. A hole is drilled through the drywall of closet A right behind the outlet box at point A. A fish tape or coat hanger is used to fish the cable from the outlet box at point A through the hole into closet A. The cable is run from the screw terminals of the jack at point A into closet A, down the wall to the baseboard, around the baseboard and thru the wall into closet B, along the baseboard of closet B, and up the wall to the desired placement for point B. A hole is cut in the wall at point B to accept the round modular jack and the mounting ring screwed to the wall using the plastic anchors. The cable is screwed to the terminals of the round modular jack and the jack and faceplate are attached to the mounting ring. This provides a neat clean new extension jack without a lot of exposed wiring.

A wall telephone is desired at point B in a hall in *Figure 5-5b*. There is a modular jack mounted in the outlet at point A. Cable is run from the screw terminals of the jack at point A along the baseboard into the lower corner of the closet. A hole is drilled at the baseboard through the wall into the closet. The cable continues to run along the baseboard, then up along the wall to the wall telephone location. The wall telephone is mounted according to ROW steps 7 through 15 of Chapter 4. The back plate of the wall

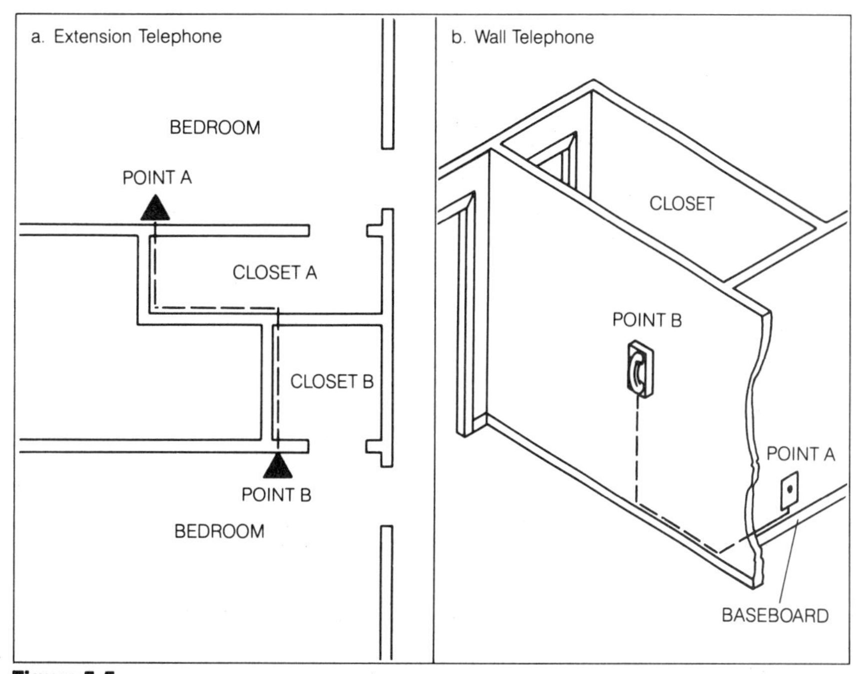

Figure 5-5.
Closet Runs

telephone modular plate (*Figure 4-15*) in this case is to be mounted directly to the wall. Position the wall telephone so that the mounting screws for the backplate screw into a wall stud; otherwise, use molly bolts to mount the backplate. Drill a hole through the wall after the position for the mounting plate has been determined, and pull the cable through for interconnection. In this manner the wall phone is mounted without any of the exposed wire being seen in the hall. Fasteners that have adhesive backing can be used to hold the cable to the inside closet wall.

CONCEALED WIRING — INTERIOR

Concealed cables are more difficult to run because they run inside of walls inside the house. In many apartment buildings it is quite difficult to run concealed wiring because there is no access to the walls from above or below. In homes and small business there may be access to the top or bottom of a wall from an attic, or a basement, or a crawl space underneath a house that is constructed with piers and beams.

Apartments With No Attic or Basement

About the only opportunity to keep the cable runs hidden besides practicing the techniques of using cabinets and closets as described in the last section, is to run the cables under the carpet along the baseboard whenever possible. When the cable runs to an outlet, it comes up from the carpet to the top of the baseboard and then enters the wall to run to the outlet. The installation is shown in *Figure 5-6b*. This provides a modular jack with only a small amount of cable exposed at the baseboard. Steps 1 through 7 should be followed for the installation.

Step	*Instruction*
1.	Locate the position of the modular jack. The one shown in *Figure 5-6* is for a round modular jack. Avoid studs in the wall; locate the position of the jack between them. Cut a 1⅜″ hole in the drywall according to the mounting instructions with the round modular jack.
2.	Directly below this hole drill a ¼″ hole at the top of the baseboard (See *Figure 5-6a*).
3.	Mount plastic anchors for the modular jack mounting ring and screw the mounting ring in place.
4.	Tie a nail or screw to a string and drop it through the hole for the modular jack. With a fish tape or stiff wire with a hook on the end, fish the string out the bottom ¼″ hole (See *Figure 5-6a*).
5.	Tie the cable wires to the string and tape the ends with masking tape to make it a smooth tapered end no larger than the cable body.

6. Pull the cable through the ¼″ hole and up through the modular jack hole. Strip the insulation from the ends of the cable (*Figure 5-6b*) and attach the wires to the modular jack body terminals with the screws provided — red to red, green to green, yellow to yellow, black to black. Mount the modular jack body to the mounting ring.
7. Put the faceplate on the modular jack and caulk around the cable in the bottom ¼″ hole. This finishes the installation.

Down Through the Attic

Take these precautionary measures when working in an attic:

1. Always wear long sleeves and protective clothing when working near insulation.
2. If flooring is not available, step only on ceiling joists. Ceilings or other materials between joists are not made to carry a great deal of weight.

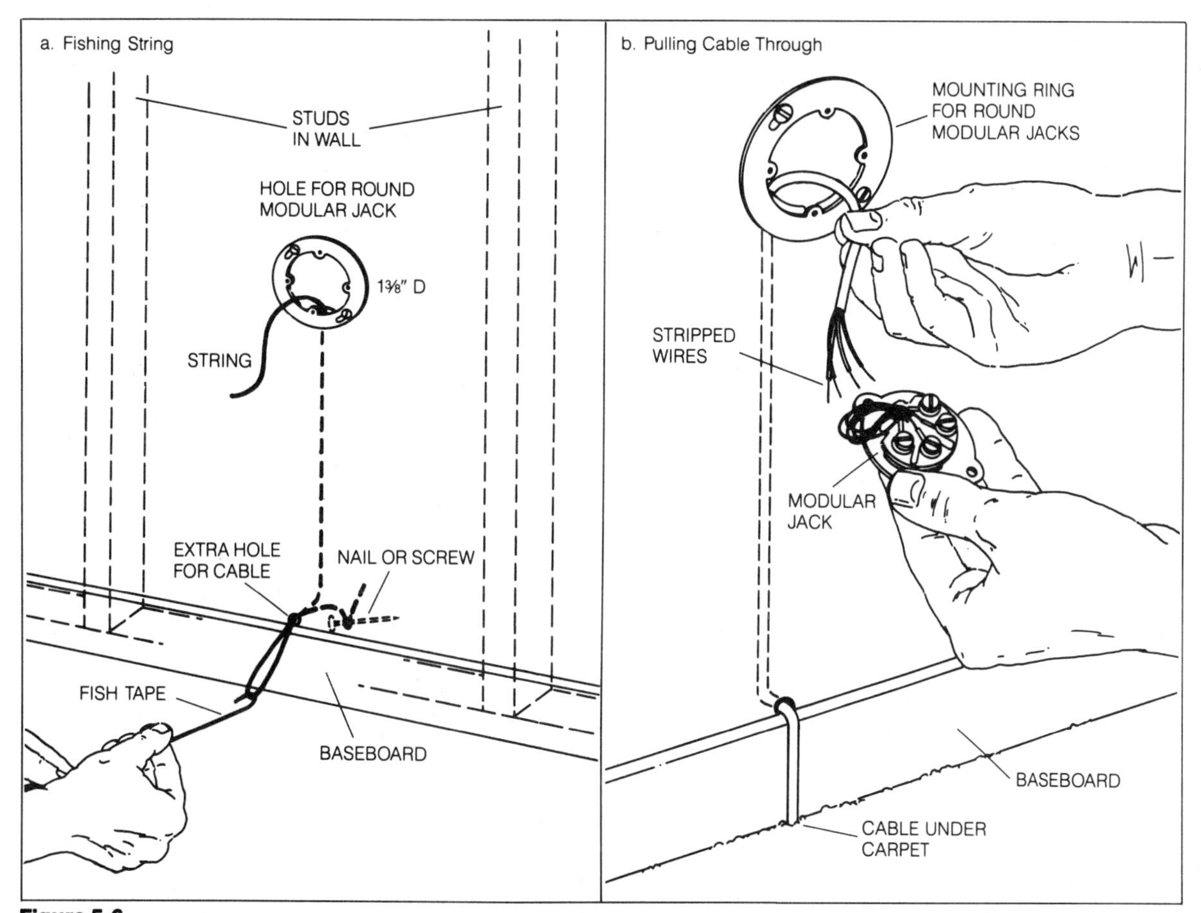

Figure 5-6.
Round Modular Jack Installation

If space is available, the attic probably is the best access for distributing concealed cables for adding telephones. The biggest problem is fishing the cable through the wall and ceiling spaces. If the wall surface were removed as shown in *Figure 5-7*, you would see 2″ × 4″ studs approximately 16″ to 24″ on centers running from floor to ceiling. At the floor level, there is one 2″ × 4″ laid on its side for a bottom plate; at the ceiling level, there are two 2″ × 4″ studs laid on their side for the top plate. Some of the studs may have fire blocks between them as shown in *Figure 5-8*.

Figure 5-7 shows the top of the wall structure from the attic. The insert shows how holes must be drilled in the top plate in order to run the cable into the wall at the correct location. This requires an 18″ drill bit of at least ¼″ and preferrably ⅜″.

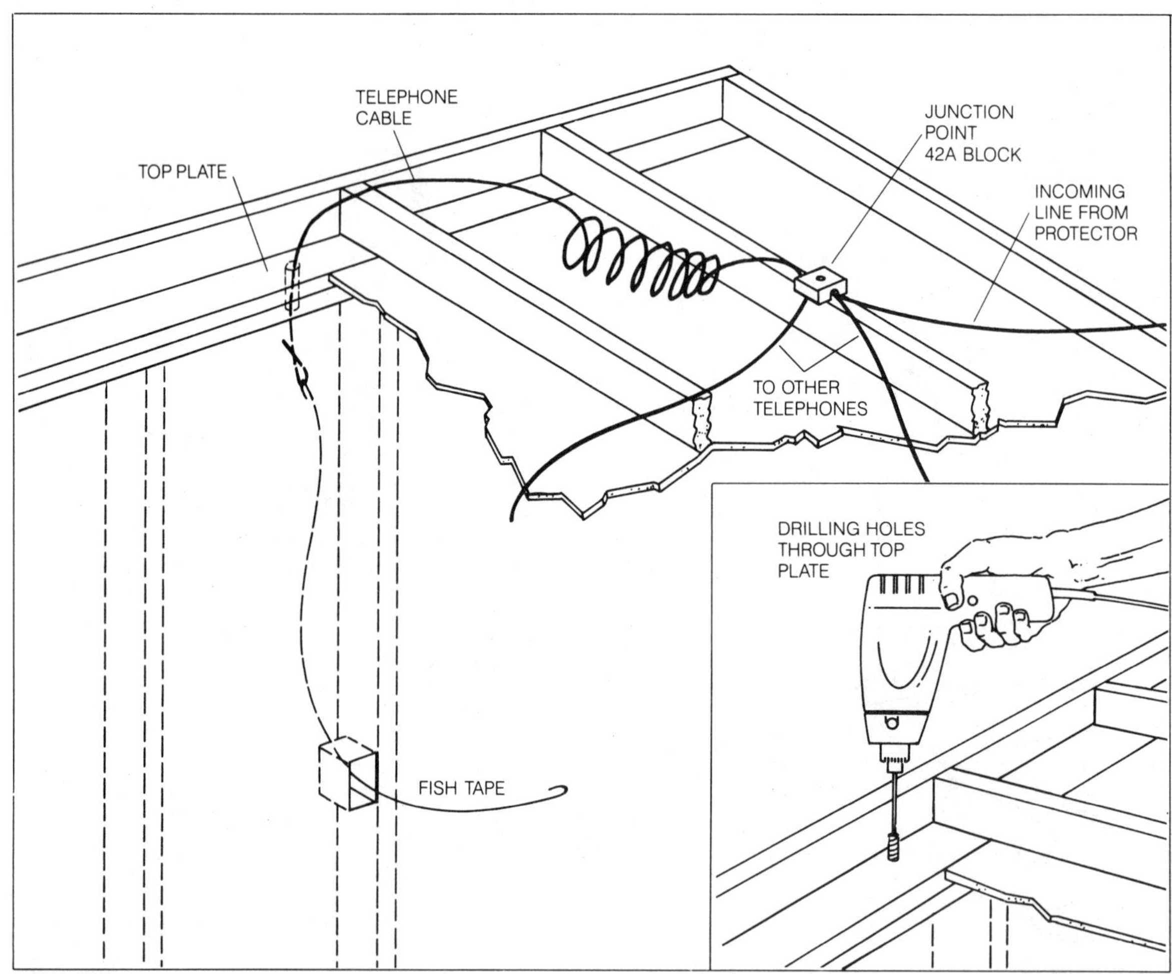

Figure 5-7.
Drilling Top Plate Hole and Pulling Cable

Here's how a typical installation would be made. Refer to *Figure 5-7* and *5-8*. A new outlet is to be installed at point A. Follow these steps:

Step	*Instruction*
1.	Determine the approximate position desired for the outlet box to hold the modular jack. In this case a rectangular modular jack is to be installed. The outlets usually are 12″ to 18″ from the floor.
2.	Locate the studs in the wall. For drywall construction, tapping on the wall will usually locate the studs. A hollow sound means no stud; a solid sound indicates a stud. When a stud has been located near the approximate position desired, drive a small finishing nail through the wall at the baseboard at points side by side until the stud positively is located. The stud need not be located to support an outlet in plaster and lathe walls, but must be located so the stud will not interfere with where the outlet is to be positioned. The easiest way to locate a stud is probably by measurement and then testing with the nail in the baseboard area. Removing the baseboard is another alternative. If one is available, an easy way to locate studs is with a magnetic nail locator.
3.	Go into the attic and locate the top plate and correct studs as shown in *Figure 5-7*. Drill a hole through the top plate with a 3/8″ drill (or at least 1/4″).

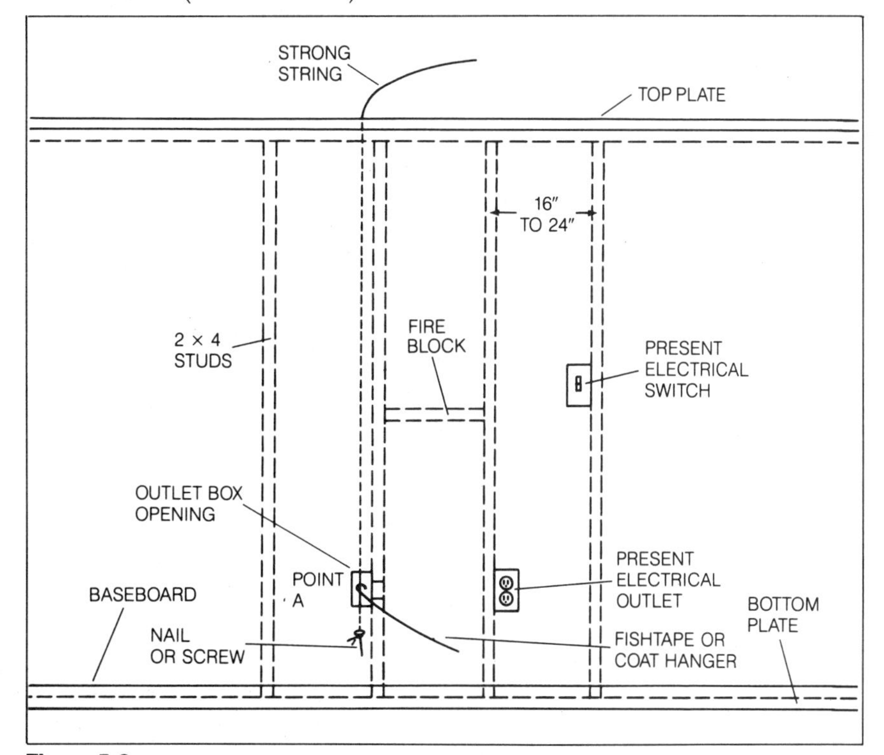

Figure 5-8.
Locating Outlet and Fishing Weighted String

4. As shown in *Figure 5-8*, tie a nail or screw or small plumbob to a strong string and drop the weighted string through the top plate hole. Have a partner listen in the room where the outlet box is to be installed. If a noise is heard at the floor level when the weight is bounced up and down, the wall is clear and you can proceed to step 5. If the weight is stuck at the fireblock level of *Figure 5-8*, the location of the outlet may have to be moved over a stud so that the fireblock can be avoided. This requires locating a new stud and drilling a new hole through the top plate.
5. With a clear path established in the wall, make a template from the dimensions of *Figure 5-9a* and cut the hole for the outlet box. Note the box to be used for drywall is one with a bracket on the side that can be nailed into the face of the stud. For plaster and lathe walls, chip the plaster away first then cut the hole in the lathe.
6. After the hole is cut, fish the string through the outlet hole with a fish tape or a coat hanger that has been straightened out and a hook put in the end as shown in *Figure 5-8*.
7. One of two methods can be used to get the cable into the wall. In the first method, pull a fish tape up through the top plate hole by attaching it to the string at the outlet box. Use the fishtape to pull the cable back down from the attic through the top plate as shown in *Figure 5-7*. In the second method, tie the cable to the string at the outlet box and pull the cable up through the top plate into the attic.

When pulling the cable up with the string, make the end of the cable tapered with masking tape so it will come through the top plate hole easily. Otherwise the string will break and you'll have to start over.

8. Pull the cable through the outlet box and mount the outlet box as shown in *Figure 5-9a*.
9. Strip about 3″ of the overall cable insulation and about ½″ of the insulation from the four conductors and attach the bare wires to the modular jack as shown in *5-9b* — red to red, green to green, yellow to yellow and black to black.
10. Mount the modular jack bracket in the outlet box with the two long screws provided.
11. Mount the faceplate to the bracket to provide a flush-mounted modular jack.

Attic Distribution

As shown in *Figure 5-7*, a 42A block can be used as a terminal strip in the attic. If this is located at a convenient place in the attic, it becomes the junction point that was discussed in *Figure 5-2*. The 42A has a quick connect modular jack connected to it with spade lugs. A telephone wiring block of *Figure 5-2* is plugged in the modular jack, and the interconnecting cables wired to the telephone wiring block to make the quick disconnect system.

Tight Places In Attic

When doing attic installations, if holes are going to be drilled through the ceiling into the attic, make sure there is clearance to get to the hole and also that there are no electrical power lines in the way. Some walls are very difficult to get to because of the roof line. Check this out before any holes are drilled.

Going Down Exterior Walls

Most of the installations are to be in interior walls. Exterior walls are very difficult to get down because of insulation. Unless there are wires in the walls already that can be used to pull up or down other wires, going down exterior walls should be avoided. It also is very difficult to get to the exterior walls because of the roof line.

Up Through the Floor

As shown in *Figure 2-5,* in many cases the telephone line comes into a basement of a house and connects to a protector in the basement. This provides an excellent opportunity to take off with a new installation by having the quick disconnect in the basement right close to the protector. As shown in *Figure 5-10,* wires from the protector can go to a 42A block that has been converted to a modular jack by one of the means discussed in *Figure 4-4* to *Figure 4-6.* A telephone wiring block plugs in the modular jack of the 42A cover to be the quick disconnect junction point.

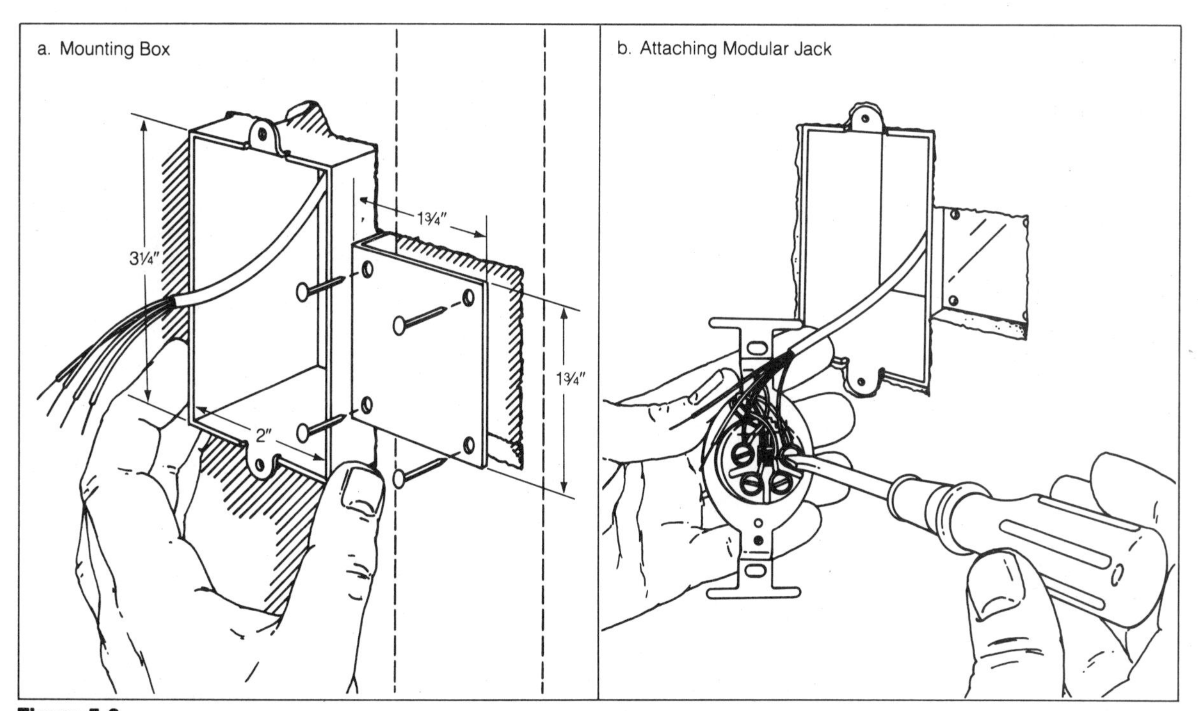

Figure 5-9.
Mounting Outlet Box and Modular Jack

Installing An Outlet — Up Through the Floor

Installation of a new outlet using a cable coming up through the floor is essentially the same as the previous steps for the attic except the holes are drilled through the floor and through the bottom plate as shown in *Figure 5-10.* Actually the basement (or this could apply to the pier and beam construction as well) is somewhat simpler because the run to the outlet usually is shorter, and there is not as much chance of a fire block being in the way.

Locating the correct stud, drilling, cutting the outlet opening, fishing the cable, mounting the box, the jack and the faceplate are the same as the attic installation.

If the protector is on the outside and the telephone line from the protector comes into the basement (or under the house), the same type fan-out as shown in *Figure 5-10* can occur.

An Easier Way From the Attic

In very difficult cases, or installations where you are willing to compromise on the concealed wiring a bit, there is an easier way from the attic. It is shown in *Figure 5-11.*

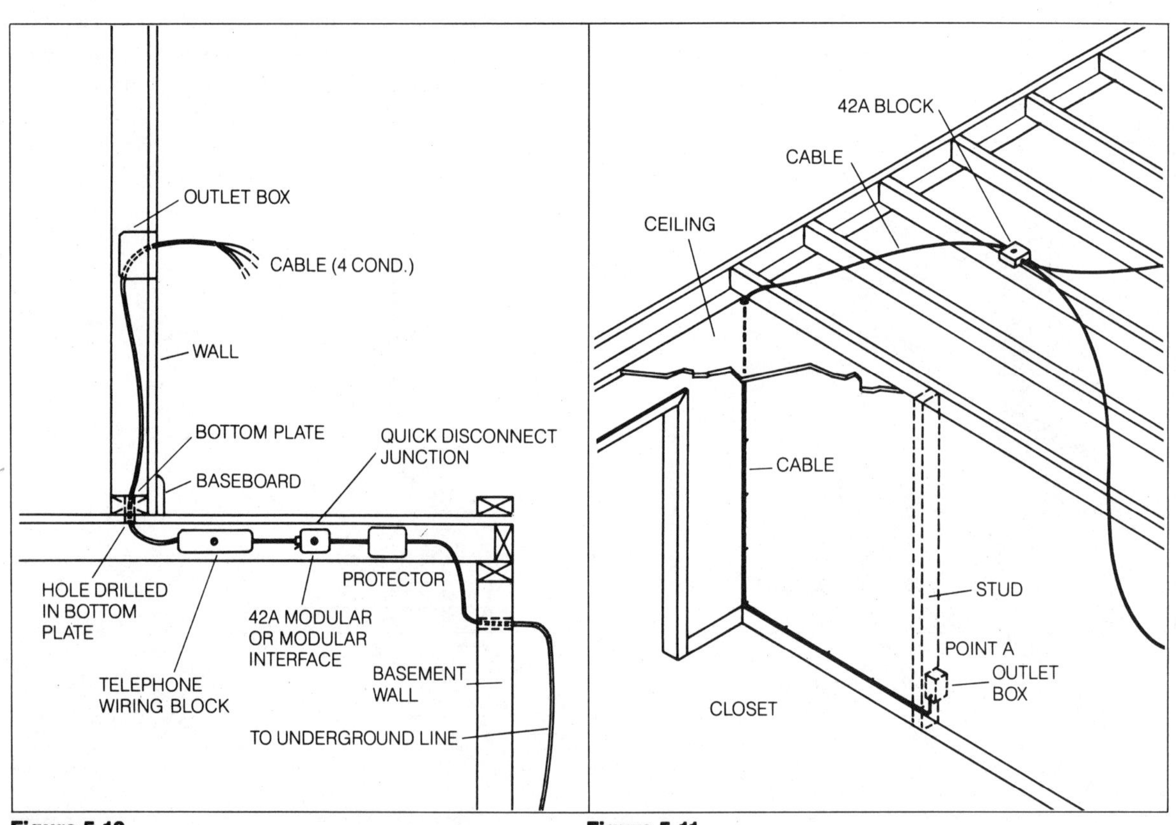

Figure 5-10.
Basement Distribution with Quick Disconnect

Figure 5-11.
Down From Attic Through Ceiling

An outlet box is to be installed in a room at point A. Behind the outlet box wall is a closet. Instead of running the cable for the outlet box inside the wall, a hole is drilled in the ceiling right at the corner of the closet. The cable is run through this hole, down the corner of the closet, along the baseboard of the closet wall, up the wall just behind the outlet box, and through the closet wall into the outlet box on the other side of the wall. In this fashion, there is no need for drilling through the top plate, fishing the cable and pulling the cable. The wiring is not concealed but it is hidden in the closet.

Repairs

Any holes cut in the wall many times do not match the outlet exactly; therefore, some repairs are needed. Patching dry wall construction can be done easily with a quick plaster used for patching. This can be applied, dried, sanded and painted in less than two days.

EXTERIOR WIRING

For particular kinds of house construction, there may be a need to run more than one exterior cable from the protector to different parts of the house. This may be because of cathedral ceilings that block access, architectural separation of parts of the house, etc.

If there is a need for you to add a cable to the protector and run it on the exterior of the house to a new entry inside the house, check your local telephone company to see if they will allow you to make the connection to the protector. Many companies are concerned that the protector is going to be damaged and as a result, lightning will damage your house and telephone facilities. Check with them to be sure.

If you run exterior wire, try to run the cable in protected places, like under the soffitt, under the edge of siding, or under the edge of a trailer. Use drip loops so that moisture will run away from the protector or away from where the cable enters the exterior wall. Use anchors for mortar or brick runs and good quality insulated staples for wood runs. Caulk all openings to protect against moisture and insects.

After locating where the cable is to enter the house through the exterior wall, check the inside and outside position carefully before drilling to make sure there are no electrical, water or gas lines near where the drilling is to occur. Insert a straightened coat hanger into the drilled hole to make it easy to pull through the cable. Caulk the hole after the cable is in position.

TRAILERS

The point from which telephone service is wired to a trailer, as with a home, is the protector. The protector either is mounted on a separate short pole called a trailer stake, as shown in *Figure 5-12,* or it is mounted on a regular pole that is bringing utility power to the trailer. The company telephone line either runs underground from a cable termination box to the trailer stake and then underground to the trailer; or overhead to the power pole, down to the protector and underground to the trailer. Of course, there can be many variations, but there should be some similarity to the above cases.

The best way to run the cable for adding or changing telephones in a trailer is to work from underneath. This corresponds to a basement distribution in a home or apartment. Never drill through the wall of a trailer. Drill through the floor, preferably near an outside wall because wires and plumbing normally are not run along the outer edge of a trailer. On the other hand, when walls are located and telephones positions are established near the middle of the trailer, be very careful drilling because usually there will be wiring and plumbing in these locations.

Follow the steps outlined for installations in the home or an apartment for the trailer installations. The line from the protector may very well terminate at a 42A block before it is distributed underneath to the existing telephone(s). Use the same 42A block for any added telephones or mount a new 42A block to start the new installations.

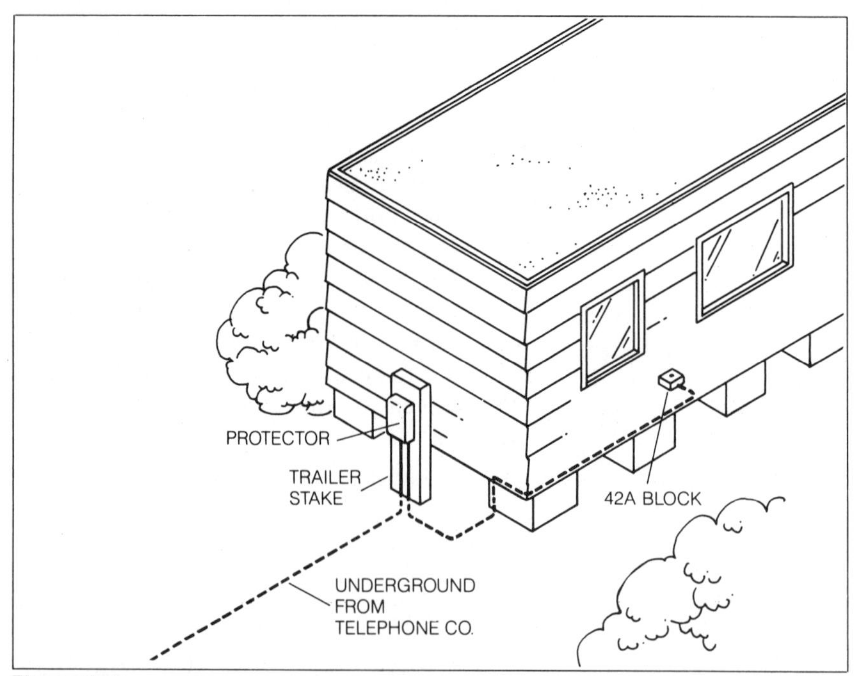

Figure 5-12.
Trailer Distribution

Replace existing telephones by following the appropriate instructions for modular and old style systems. Add telephones using the instructions for modular or old style systems as they apply to your situation. The major difference is likely to be the type materials that are used in the trailer compared to the home or apartment. In the trailer there will be finished interiors of metal, plastic and moulded fiberglass, while in a home or apartment it is wood, drywall or plaster and stone or brick.

If it is necessary to run a new cable from the protector, run it underneath to the trailer. At the trailer, keep the new cable hidden and protected as much as possible by running it along the underneath edge of the trailer until it needs to be turned in to run to a 42A junction point. From the 42A block, distribute it to the point where it must go through the floor and into a wall, closet or cabinet to the newly added outlet. In many cases, drilling down through the floor is required. Carefully identify the position of the hole to be drilled so that everything is clear underneath before the hole is drilled. Make the final faceplate mountings and telephone set installations as you would in any home or apartment.

PULSE
TONE
1
ABC 2
DEF 3
GHI 4
JKL 5
MNO 6
PRS 7
TUV 8
WXY 9
REDL
*
OPER 0
#
HOLD

6 ADDING TELEPHONES TO A MODULAR SYSTEM

Chapter Contents	*Page*
Case History #1	
Simplified Technique	*2*
Case History #2	
No Attic or	
Basement Access	*5*
Case History #3	
Basement or	
Through Floor Access	*8*
Multiple-Line Telephones	*12*
Case History #4	*12*
2-Line Telephones	*13*
Connections Required	*13*
Are Connections in Place?	*15*
Area Codes	*18*

Your present apartment or house has a modular telephone system installed. Besides replacing your existing telephones, there is a need to add telephones. This chapter explains how to do such installations.

After reading this chapter you should:

A. Know some techniques for adding telephones to an apartment that has no attic or basement access.

B. Know some techniques for adding telephones to an apartment or house with basement access. Attic access, if available, is covered in Chapter 7.

The techniques for adding telephones will be demonstrated by case histories.

CASE HISTORY #1 - SIMPLE TECHNIQUES

Figure 6-1 is the same apartment floor plan as shown in *Figure 3-1*. However, more detail has been added. The cabinets and closets are located so they may be used effectively if required to conceal cable installed for the added telephones. Also, the present location of the bed in the master bedroom is shown.

Several rearrangements are under consideration. The first is to add a desk, as shown by dotted lines, in the master bedroom and an outlet at A for a telephone on the desk. The bed would be moved to a new location, as shown by dotted lines, and an outlet installed at B to connect the extension telephone by the bed. Such an installation is quite simple, follow these steps:

Step	*Instruction*
1.	Plan the installation, make a bill of materials and purchase the parts. In this case, it was decided to use a modular duplex jack as shown in *Figure 6-2a* at the present outlet P, and run cable to two surface-mounted jacks shown in *Figure 6-2b* to be screwed to the baseboard at A and B. Extension cords shown in *Figure 6-2c* that have a modular plug on one end and spade lugs on the other will be used as cables to connect the outlets at A and B. Two are required.

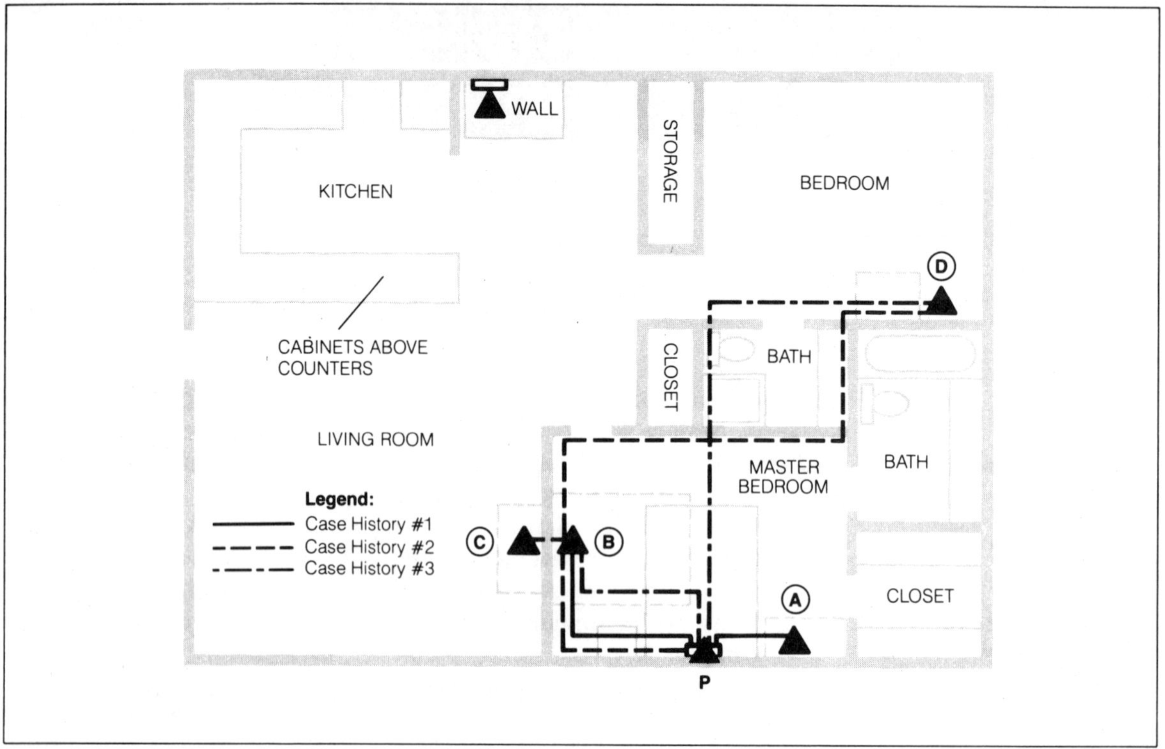

Figure 6-1.
Apartment Floor Plan

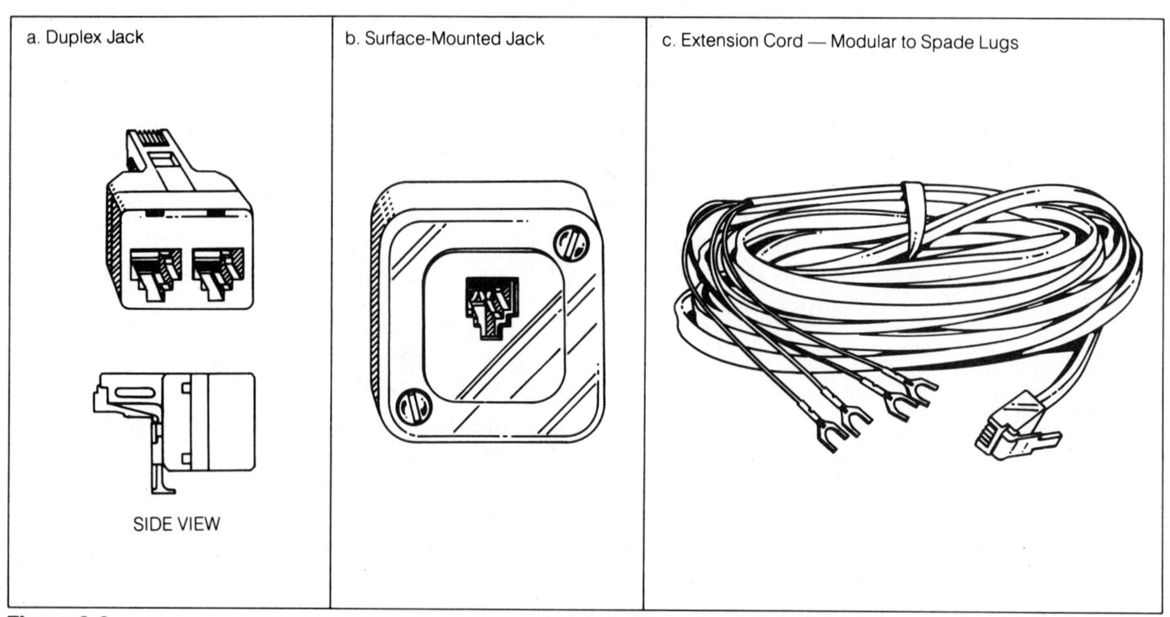

Figure 6-2.
Installation Parts for Case History #1

The installation is shown in *Figure 6-3*.

2. Locate the position for the surface mounted jacks A and B.
3. Plug the duplex jack in the outlet at P.
4. Connect the modular plug of one of the extension cords to one modular jack of the duplex jack. Dress the cord down to the baseboard and run the cord along the top of the baseboard and staple it with an insulated staple at this point (initial staple). Dress the cable all along the baseboard to the position for outlet A. Now unplug the duplex jack.
5. Connect the spade lugs to the surface-mounted jack as shown in *Figure 3-3*. If the cord is too long, cut it to the correct length, and strip the outer cable and wire insulation to expose bare wires (See *Figure 3-4*). Connect the bare wires to the surface-mounted jack — red to red, green to green, yellow to yellow, black to black.
6. Break away a slot for the cord to run out the top of the jack as shown in *Figure 3-3* and screw the surface-mounted jack to the baseboard. Staple the cord to the baseboard every 3 or 4 inches.
7. Repeat steps 3 through 6 for outlet B.
8. Plug the duplex jack in outlet P and plug in the telephones.

The installation is complete. The cord could have been run under the carpet as shown by the dotted lines in *Figure 6-3*. Less of the cord would be exposed and the installation might be a bit neater. Remember to run the cord under the carpet right next to the baseboard so there will not be a chance of wear due to walking on the cord (See *Figure 5-3*).

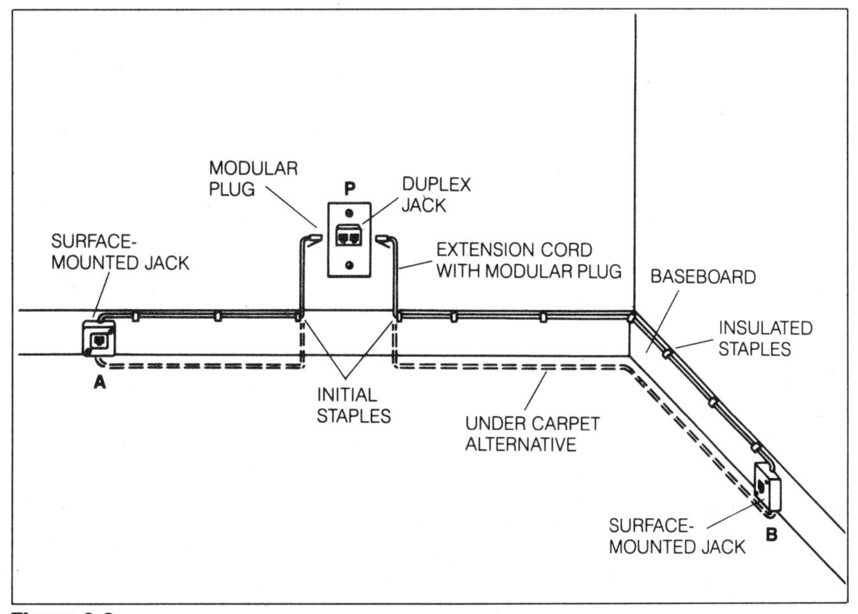

Figure 6-3.
Simple Installation for Outlets at A and B for Case History #1

CASE HISTORY #2 - NO ATTIC OR BASEMENT ACCESS

A second rearrangement of *Figure 6-1* moves the bed to the new location indicated by dotted lines, but the desk is not installed in the master bedroom. It is installed in the living room as indicated by the dotted lines. As a result, an outlet at B and C are required. In addition, a telephone is to be added for the children in the back bedroom. An outlet D is required for this telephone. Here is how the outlets are installed:

Step	*Instruction*
1.	Plan the installation by plotting the cable runs on the floor plan (As shown on *Figure 6-1*). Make a bill of materials and purchase the parts. In this case an extension cord shown in *Figure 6-2c* again is run from P to the outlet B. No duplex jack is required at P. At B, a 42A block is used as a junction point for cable runs to outlet C and D. It is mounted on the baseboard and has a quick-connect jack as a cover which provides the modular jack for outlet B. The quick-connect jack may connect to the 42A block with spade lugs or snap-on connectors (See *Figure 2-6* and *4-5*). A cable goes through the wall right at the baseboard to make the connection to a 42A block for outlet C which is the same type outlet as B. The installation is shown in *Figure 6-4*. The cable for outlet D runs from the 42A block at B along the baseboard, around the doorframe, along the baseboard, through the wall to the sink cabinet in the bathroom, through the bathroom inside the sink cabinet, through the cabinet and wall into the bedroom,

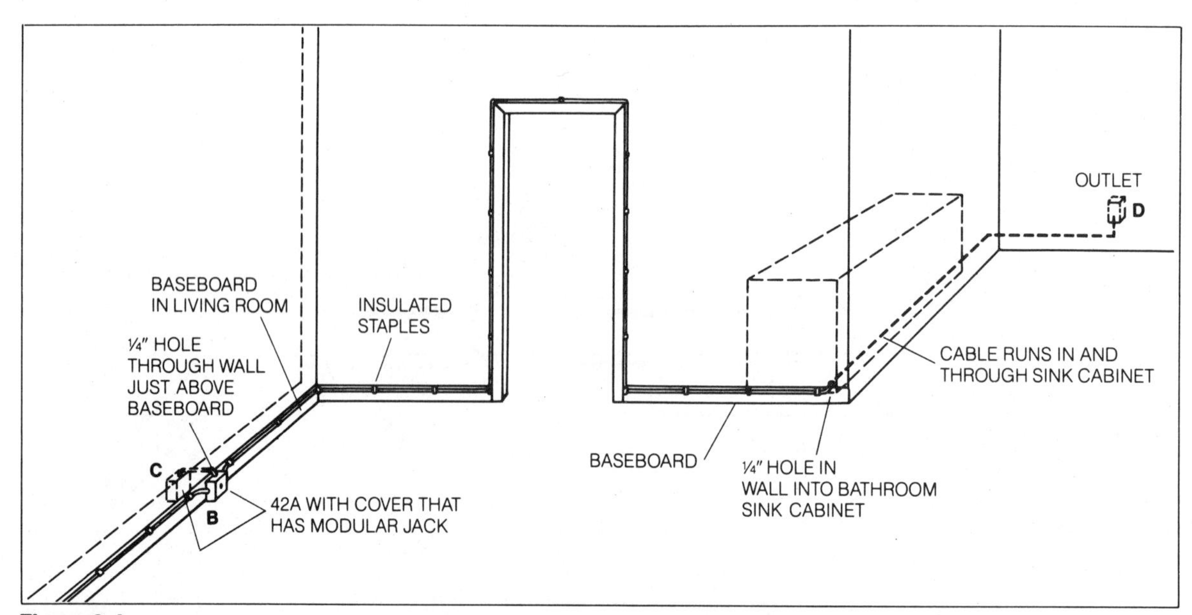

Figure 6-4.
Installation for Outlets B, C and D for Case History #2

along the baseboard to the position of outlet D, into the wall at the baseboard, and to an outlet box for a rectangular modular jack for outlet D (See *Figure 2-10a)*.

2. Locate the position for the 42A blocks of outlets B and C and the outlet D box. Outlet D was chosen to have a rectangular outlet box like that shown in *Figure 5-9*. It could have been a round outlet as shown in *Figure 2-10a* and *Figure 5-6*.
3. Connect the modular plug of the extension cord to outlet P. Dress the cord down to the baseboard and run the cord along the top of the baseboard and staple at the initial point. Dress the cable all along the baseboard to the position of outlet B. Now unplug the cord from outlet P. This prevents contacting electricity in case the present telephone should ring.
4. Connect the spade lugs to a 42A block. If the cord is too long, cut it to the correct length, strip the insulation (See *Figure 3-4*), and connect the base wires to the screw terminals — red to red, green to green, yellow to yellow and black to black.
5. Mount the 42A block for outlet B to the baseboard.
6. Drill a ¼″ hole through the master bedroom wall to the living room right above the baseboard. It should come out slightly above the baseboard in the living room (*Figure 6-4*).
7. Mount the 42A block for outlet C on the living room baseboard right below the hole drilled in the wall.
8. Cut a length of cable (4 conductor) and insert it into the ¼″ hole in the wall in the master bedroom. Push it out through the ¼″ hole in the wall in the living room. Leave about 6″ of cable exposed out the wall on each end. Strip the insulation from both ends and connect outlet C to outlet B by connecting the bare wires to the screw terminals of the 42A block (See *Figure 3-4*). Match the color of the conductors at each 42A block. Push any excess cable into the wall.
9. Connect and mount the quick-connect modular jack as a cover over the 42A block for outlet C (See *Figure 2-6c*).
10. Drill a ¼″ hole through the master bedroom wall and bathroom cabinet right above the baseboard as shown in *Figure 6-4*.
11. Drill another ¼″ hole through the end of the sink cabinet and through the back bedroom wall. Do this from the children's bedroom directly above the baseboard. Be careful not to drill into any plumbing.
12. Position the box and cut the hole in the wall for the outlet box as shown in *Figure 5-9*.
13. Drill a ¼″ hole below the box opening directly above the baseboard as shown in *Figure 6-5a*.
14. Push the cable through the master bedroom wall, through the bathroom cabinet and into the back bedroom. Dress it along the baseboard to the outlet D position.

15. Fish a string from the hole for outlet D through the ¼″ hole at the baseboard below the outlet D opening. Attach the cable and pull it up through outlet D opening (See *Figure 6-5*). Leave about 12″ of cable exposed from the opening.
16. Mount the box, pull the cable through the box, attach the wires to the rectangular modular jack, and mount the jack as shown in *Figure 5-9*. After the faceplate is attached, outlet D looks like the rectangular outlets of *Figure 2-10a*.
17. Dress the cable along the baseboard in the children's bedroom by pushing the slack to the master bedroom along the baseboard, around the doorframe, and back to the 42A at outlet B.
18. Cut the cable to length, strip the insulation and attach the bare wires to the terminals of the 42A block of outlet B. Keep the correct color conductors tied together.
19. Attach the terminals of the quick-connect jack to the terminals of the 42A block. As mentioned earlier, these will either be spade lugs or snap-on connectors. The 42A block at outlet B is a junction point for four lines — one to outlet C, one to outlet D, one to the quick-connect jack of outlet B and one from the outlet P.

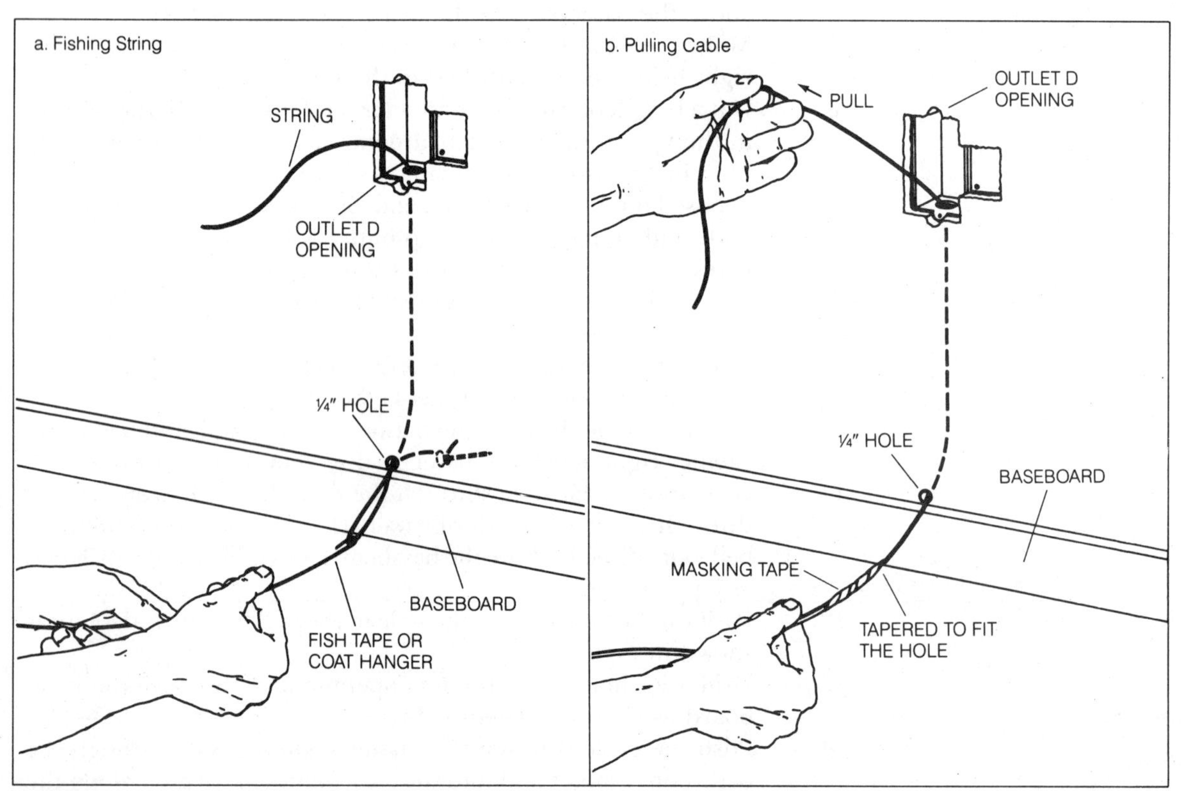

Figure 6-5.
Pulling Cable for Outlet D Case History #2

20. Put the quick-disconnect jack cover on the 42A block to complete outlet B.
21. Staple the cable down to the baseboard at all locations every 3-4 inches.
22. Connect the modular plug to the P outlet and the installation is complete. Telephones can now be plugged into outlets B, C and D.

CASE HISTORY #3 - BASEMENT OR THROUGH FLOOR ACCESS

As a comparison of installations, let's assume that the apartment of *Figure 6-1* has a basement and there is access from underneath the floor (apartments constructed on piers and beams or apartments over garages might have a similar access). Following are typical kinds of steps one would go through for this type installation.

Steps *Instructions*

1. Plan the installation by plotting the cable runs on *Figure 6-1*. Make a bill of materials and purchase the parts. In this case cables are run from outlet P to the other outlets by drilling a hole through the bottom plate of the wall and inserting the cables into the wall from underneath. The modular jack in outlet P is used as a junction point for two cables (X and Y) that run under the floor, one to outlets B and C (X) and one to outlet D (Y). All outlets are rectangular modular jacks mounted in electrical outlet boxes that have the special tab for mounting in drywall after the drywall is in place (See *Figure 5-9*). The cables run along the floor joists underneath as far as they can and then cross the joists to get to the outlet position.

An illustration of the cable runs is shown in *Figure 6-6*. Here are the installation steps.

2. Outlet B, C and D will be located on studs in the wall. Locate the wall stud and box position by following the steps and discussion for *Figure 5-8*. Outlet C will be on the same stud as outlet B but on the opposite wall. Measure from the existing walls to determine the location of outlets B, C and outlet D.
3. Go in the basement or underneath the floor and locate the position of the outlets using the inside dimensions of Step 2. Remember to add one-half the thickness of the walls (usually drywall construction has a thickness of about 4″). Make sure that everything is clear for drilling a hole through the bottom plate. If there are obstructions, move the location of the outlet to another stud where the way is clear.
4. With a partner on the inside, verify that the outlet locations are satisfactory and that you are in the correct position by taping on the floor at the point where you are going to drill the holes.

5. Using a ⅜″ drill with an 18″ bit as shown in *Figure 6-6*, drill a hole up through the bottom plate to insert the cables for outlets B, C, outlet P and outlet D. At the appropriate places where the cables will have to cross, drill ¼″ holes through the floor joists *(Figure 6-6)*. These holes may have to be drilled at an angle.
6. Cut the openings in the wall using a pattern with the dimensions shown in *Figure 5-9a*, or one similar for the box you are using. Drill a hole in the center of the pattern and cut the pattern opening with a keyhole saw. Otherwise, use a drill to drill holes side-by-side around the edge of the pattern and punch out the pattern.
7. Remove the faceplate and bracket holding the modular outlet from outlet P. Pull the bracket out of the outlet box so that terminals connecting the cable wires are accessible (See *Figure 6-7*).

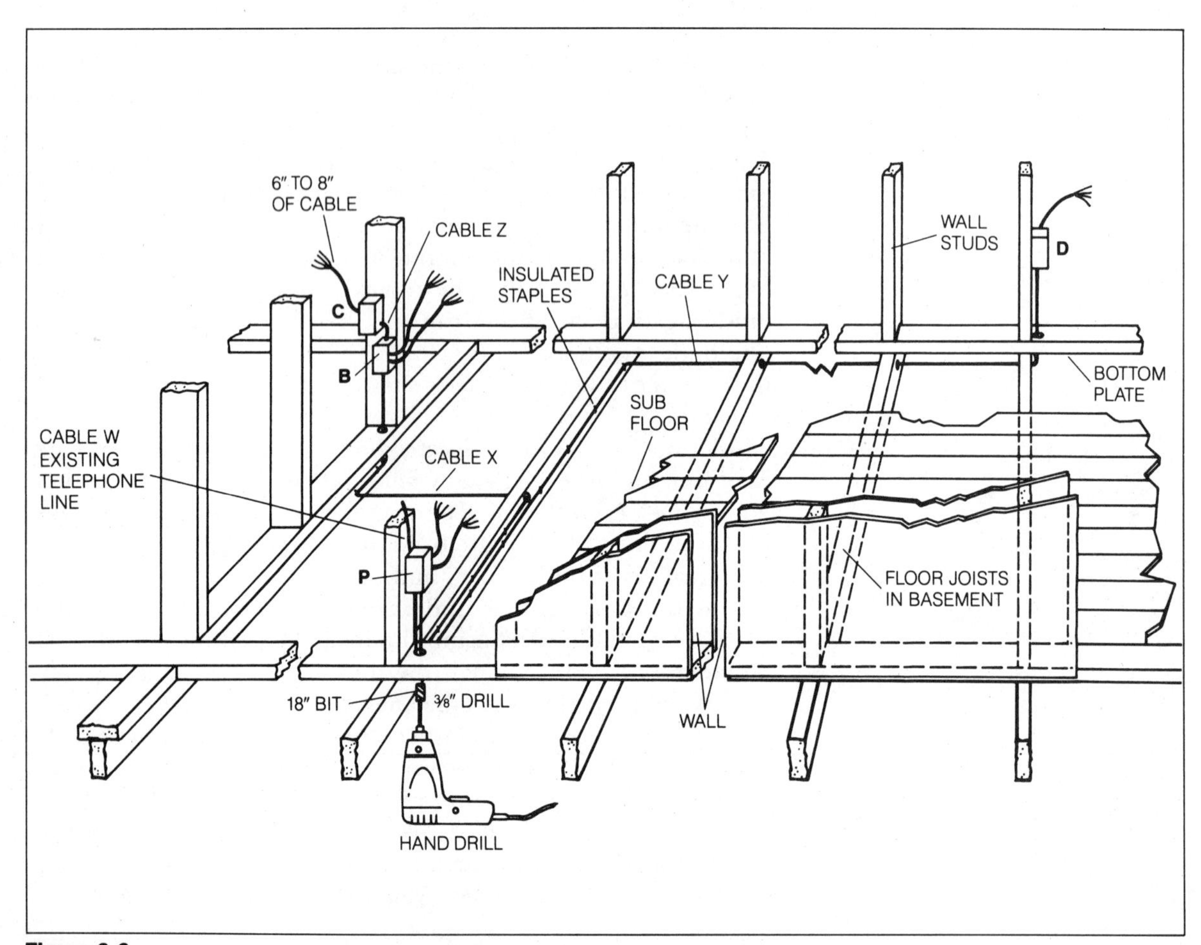

Figure 6-6.
Cable Runs Underneath Floor for Case History #3

8. Open a knockout in the bottom of the outlet P box for the cables X and Y coming from underneath. Insert a fish tape as shown in *Figure 6-7*. Have a partner put a hook in the end of a coat hanger and insert it up through the hole in the bottom plate below outlet P. Hook the coat hanger with the fish tape from outlet P and pull on the fish tape.
9. Attach the two cables of sufficient length, one for outlet B and C (X) and one for outlet D (Y), to the end of the coat hanger with masking tape. Taper the joint so it will go through the bottom-plate hole easily. Pull the cables up through the P outlet box until 6″ or 8″ of the cables are hanging out the box (*Figure 6-6*).
10. Go underneath and string the cables through the holes and along the floor joints until they are run to the holes in the bottom plate for outlet B, C and outlet D.
11. At each of the outlets B and D, fish the coat hanger and pull the cable the same way as for the cable of outlet P. (Steps 8, 9 and 10)

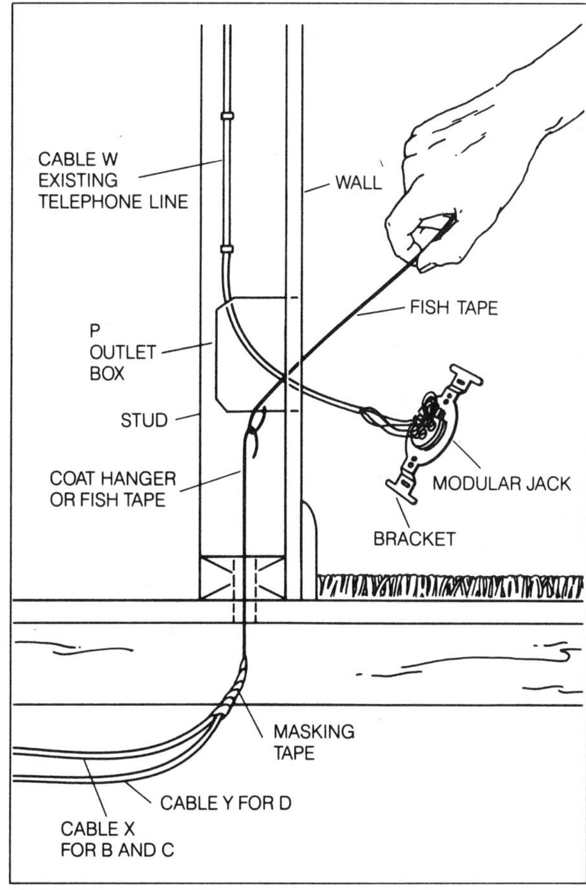

Figure 6-7.
Fishing Cable at Outlet P

12. String the cable (Z) from the opening of outlet B to the opening of outlet C. This is a short cable of about 2 feet.
13. Open the appropriate knockouts in the outlet box for outlet C and for outlet B. Insert the cable(s) into the respective box through the knockout holes. Clamp the cables in place with the box clamps (if provided) so that about 6″-8″ of cable is left extending from each box.
14. Fasten the box for outlet B to the stud and for outlet C to the stud.
15. Repeat steps 13 and 14 for the outlet box for outlet D. The cable runs should now look like *Figure 6-6* with the boxes in place and the cables extending from the boxes.
16. Connect each cable to the modular jack terminals as shown in *Figure 6-8* by stripping back the cable sheath insulation 3″ to expose the conductors and stripping the conductor insulation ½″ to expose the bare wires. Then wrap the bare wires around the screw terminals and turn the screws tight. Do outlet D (cable Y) first, then C (cable Z), then B (cable X). B has two cables (X,Z) connecting to each terminal. In all cases keep the colors corresponding—red to red, green to green, yellow to yellow and black to black.

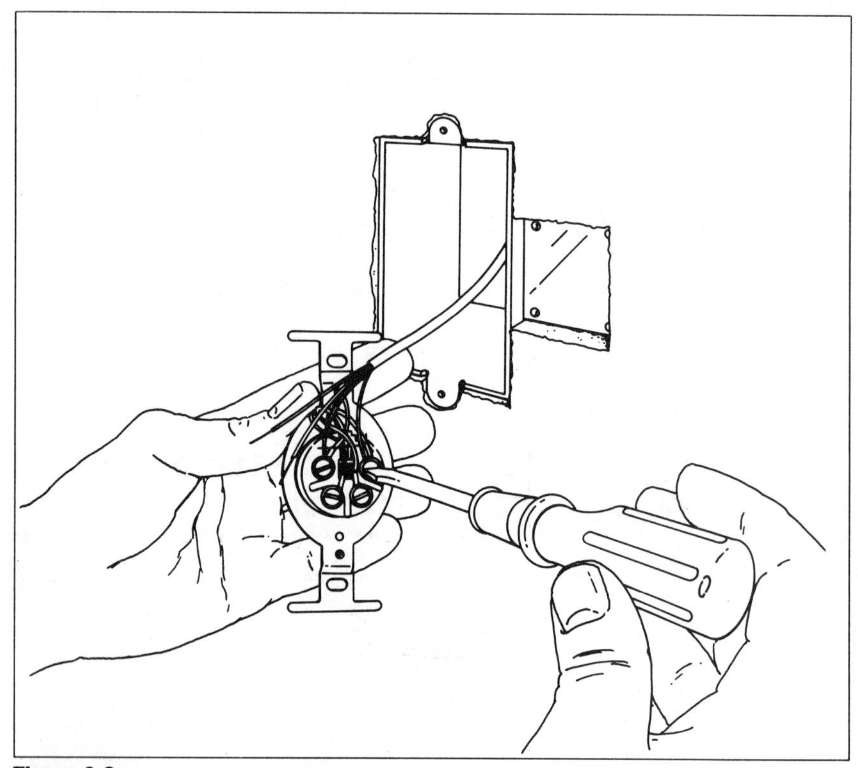

Figure 6-8.
Attaching Wires to Modular Jack

17. Before connecting cables X and Y to outlet P, lift the handset from any telephone set to go off hook and protect you from shock. Connect cables X and Y to outlet P. The modular connector at outlet P now serves as a junction point for three cables (W, X and Y). W is the incoming telephone line.
18. Test each outlets' modular jack by plugging in a telephone after the telephone handset removed in step 17 is returned on hook. When the telephone is plugged in outlet P, B, C, or D and the handset lifted, a dial tone should be present. If it is not, check all connections carefully to make sure there is no short or open connection.
19. With each outlet connection proven, install the brackets holding the modular jacks in the outlet boxes and screw on faceplates. Coil any excess cable into the outlet boxes.
20. Dress the cable in the basement along the joists and staple it with insulated staples. Push any excess cable into the walls.

The installation is complete and ready to receive the added telephones.

MULTIPLE-LINE TELEPHONES

Advances in integrated circuit functional density and reduced manufacturing costs have continued to add feature after feature to telephones available on the market. From one-piece designs with pulse-tone dialing selection, redial, lighted keypad and ringer selection; to two-piece wall or desk telephones with hold button, fashion models, one-key memory dialing, big keys for easy dialing, and amplified handsets; to full-feature wall or desk telephones with memory dialing by names as well as numbers, speakerphone for "hands-free" operation and conference calling, alphanumeric displays to show name or number called, time of call, date and time of day, and special flash, pause and hold buttons; the expansion is phenomenal.

All of the above features are available on single-line telephones. However, the family and business use of the telephone is requiring expansion of the telephone instrument so that it can be used on more lines than one, and the features are being expanded for use on the multiple lines. For this reason, we will look at a case history of installing 2-line telephones into an apartment or home.

CASE HISTORY #4 — 2-LINE TELEPHONES

Let's consider the same apartment layout as in *Figure 6-1* and assume the case history #3 arrangement of lines has been made. The family now is several years older, the children have grown and developed friends of their own, and the father is involved in a part-time business at home in the evenings. As a result, the family decided to install another telephone line.

2-Line Telephone Line

Two questions must be asked if you wish to install a 2-line telephone. First, what are the connections required? And second, are the required connections in place? Let's look at the first question?

What are the Connections Required?

The required connections for the single-line telephones presently installed in the apartment according to Case History #3 are shown in *Figure 6-1a*. The incoming line from the telephone company's protector terminates in an outlet box at point P. P has a modular outlet which is not being used so it is omitted from *Figure 6-9a,* a simplified illustration of the connections. The single active line from the telephone company terminates in the outlet box, and the red and green wires of the cable distributing the line around the apartment are connected to the line. Most of the time the incoming cable conductor colors are red and green; however, some times they are different colors (see *Table 5-1*). At each modular outlet, the red and green wires distribute the line to each single-line telephone. Notice that only two wires are required, and as you will see in *Figure 6-10*, only an RJ11 modular plug and jack are required at each outlet, and each telephone, for everything to work properly.

In most modern day installations, more connections than just the red and green wires are made at each location. An examination of the connections to the outlet box at P, and to all the modular outlets shows that the yellow and black wires have also been connected. This also is the case of the incoming cable from the telephone company. Therefore, the total connection schematic is as shown in *Figure 6-9b.* The yellow and black wires are represented with dotted lines because they are not active connections. Even though the wires coming from the telephone company are brought to outlet P, they are left open at some cable termination box and are not connected to the central office. *For a home or an apartment with a single-line telephone system, the yellow and black wires usually are inactive; however, sometimes they are used for special features — ones that we have mentioned are a lighted dial or lighted key for a wall or desk telephone, or transformer power.*

The point here is that even though only two wires are required for a single-line system, most modular outlets, plugs, jacks and line cords have all four connections provided. Therefore, as shown in *Figure 6-10*, instead of a modular plug and jack being an RJ11 with only two wires, they are really RJ14 plugs and jacks with four wires. RJ14 modular connectors are required for 2-line telephone systems; only RJ11 connectors are required for single-line systems.

The connections required for a 2-line telephone system are shown in *Figure 6-11*. The new line, line 2, is brought into the outlet box at P to connections which use the yellow and black wires for distribution throughout the apartment. Therefore, all that needs to be done to provide 2-line service to the apartment is to have the telephone company connect the incoming cable connected to the yellow and black wires at outlet P to the central office. All other connections are already in place.

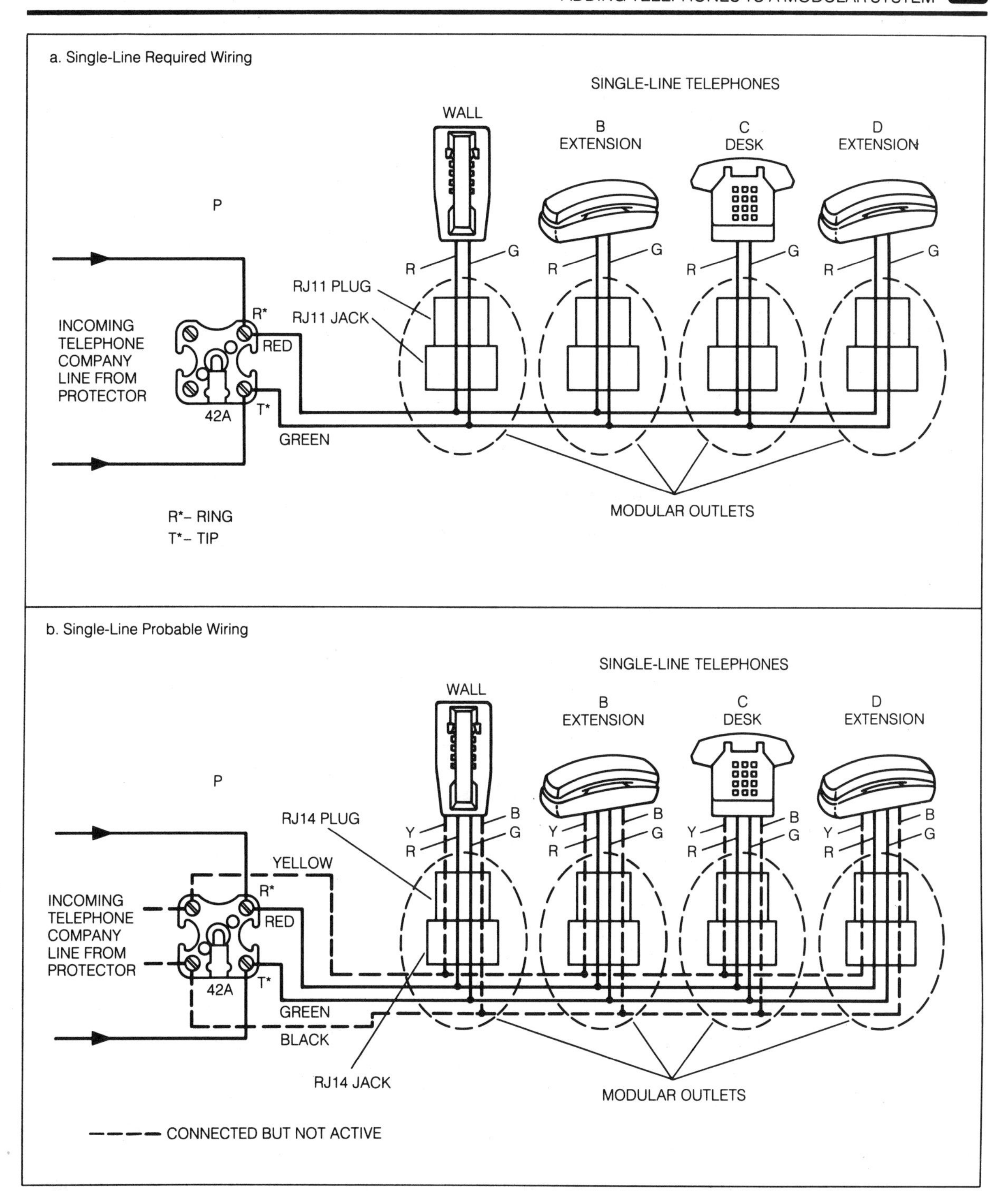

Figure 6-9.
Single-Line Wiring

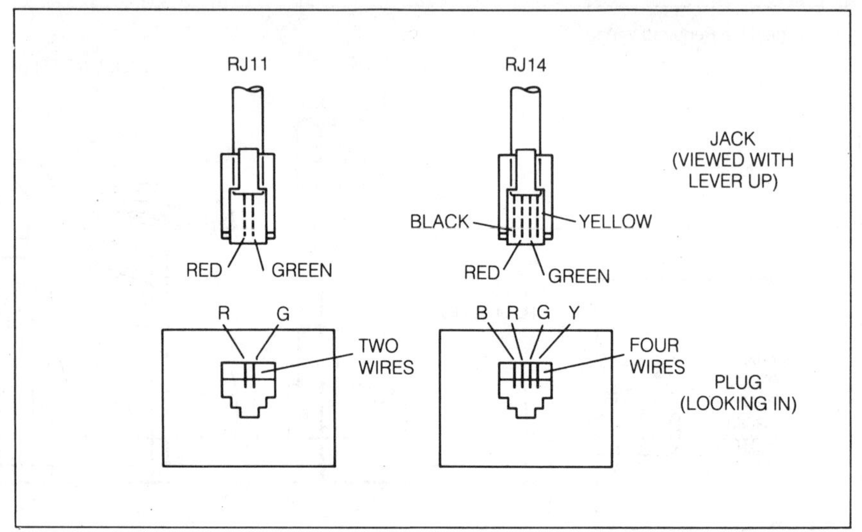

Figure 6-10.
Comparison of RJ11 and RJ14 Modular Connectors

When you order a second line from the telephone company, some telephone company installer will go to a cable junction box and make the proper line connections to provide the service.

After the service is connected, all telephones that are to use both lines will have to be 2-line telephones. Single-line telephones will still operate on Line 1 at any of the modular outlets.

Are the Required Connections in Place?

To assure that all of the connections shown in *Figure 6-11* are in place, each modular plug and jack must be examined to make sure they are RJ14 and have the 4-wire connections. Follow these steps:

Steps *Instructions*

1. Visually inspect each modular outlet and assure that the jack has the 4 wires for a RJ14 jack. This may require that the faceplate or quick-connect jack be removed to see that the red, green, yellow and black wires are all connected.
2. If some of the modular outlets are not RJ14, switch them out for RJ14s. If the connecting cables happen to have only two conductors, then new cables need to be added for the new line. The techniques used previously would be used to install the new cable runs. A 2-line coupler would be used to couple the two separate lines to the 2-line telephone. The hook-up is shown in *Figure 6-12a.*

Completing the Installation

3. Select 2-line telephones for each of the positions where operation on two lines is desired. The 2-line telephones will have RJ14 plugs on the end of their line cords. *Table 6-1* lists various types of 2-line telephones and their features.

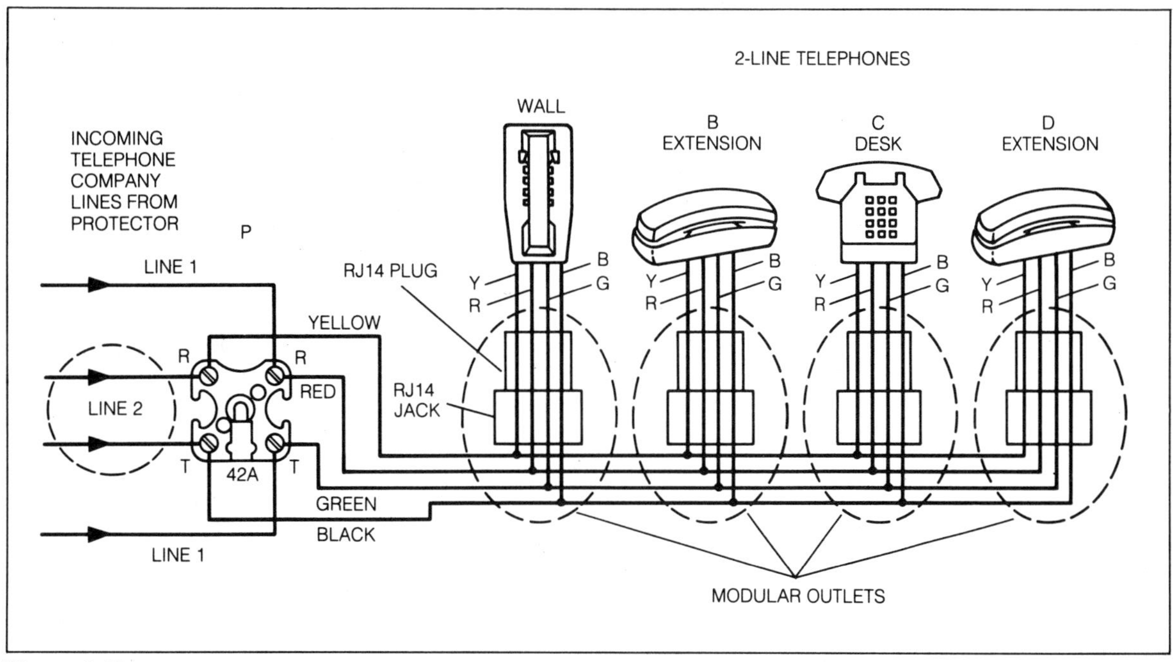

Figure 6-11.
Required 2-Line Wiring

Table 6-1.
2-Line Telephones

Type		Features*
Wall Desk	①	Line and Hold Selection Ring/Hold Indicator Touch Redial
Desk/Wall		Same as ① with Flash Button Memory Dialing
Desk/Wall Compact		Same as ① with Conference Button Separate Ring and Hold Indicators
Telephone/Intercom	②	Line and Hold Selection Line Status Indicators Touch Redial/Pause Flash Button Memory Dialing FM Intercom
Speakerphone		Same as ② Less Intercom or with Speakerphone
Full-Feature		Same as ② Less Intercom with Appointment Memory Name and Number Dialing Auto Redial (10 Times) Conference (3 Parties) Speakerphone Call Timer Date, Time Display Number Dialed Display

4. As shown in *Figure 6-12b*, it is possible to continue to use single-line telephones and distribute their use along the installed 2-line system. A 2-line coupler is required at each RJ14 modular outlet.
5. Call the telephone company and request a second line.

In summary, installing 2-line telephones can be as simple as calling the telephone company, requesting to have a second line connected, and replacing single-line telephones with 2-line telephones. If all connections are not in place, it need be no more difficult than installing new modular outlets. Replacing all single-line telephones at once maybe to expensive a task. One-at-a-time replacement is possible because single-line telephones will continue to operate on line 1 just as before. As the single-line telephones are replaced, make sure you notify the telephone company of telephone REN ratings, and keep in mind that a 5.0 REN should not be exceeded on each line.

One additional item, if you want your 2-line telephone to "roll-over" and automatically ring the idle line if the called line is busy, this feature must be ordered from the telephone company.

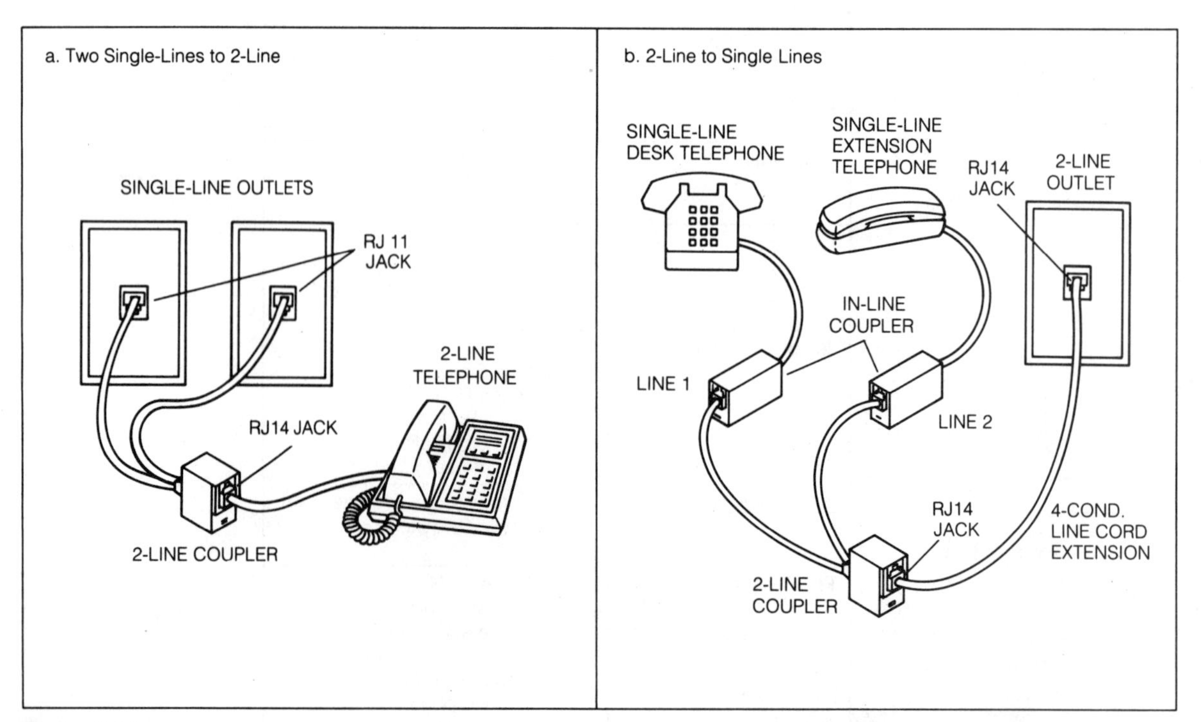

Figure 6-12.
Use of 2-Line Coupler

TELEPHONE AREA CODES

***United States + Canada + Mexico** (Dial 1 + Area Code + Local Number)*

Alabama
Birmingham 205
Mobile 334

Alaska
907

Antigua
268

Arizona
520
Phoenix 602

Arkansas
501

Bahamas
242

Barbados
248

California
562
909
Bakersfield 805
Fresno 209
Long Beach 310/562
Los Angeles 213
Oakland 510
Orange County 714
Pasadena 818
Sacramento 916
San Diego 619
San Francisco 415
San Jose 408
Santa Rosa 707

Canada
905
Alberta 403
British Columbia 250
London 519
Manitoba 204
Montreal 514
New Brunswick 506
North Bay 705
Nova Scotia 902
Ottawa 613
Prince Edward Isle. 902
Quebec, Que. 418
Saskatchewan 306
Sherbrooke 819
Thunder Bay, Ont. 807
Toronto 416
Vancouver 604

Colorado
Aspen 970
Denver 303
Pueblo 719

Connecticut
Hartford 860
Bridgeport 203

Delaware
302

District of Columbia
Washington 202

Florida
561
Fort Lauderdale 954
Fort Myers 941
Gainsville 352
Jacksonville 904
Miami 305
Orlando 407
Tampa 813

Georgia
Macon 706
Atlanta 404
Savannah 912

Hawaii
808

Idaho
208

Illinois
Centralia 618
Chicago 312/773/847/630/708
Peoria 309
Rockford 815
Springfield 217

Indiana
Evansville 812
Indianapolis 317
Lafayette 765
South Bend 219

Iowa
Council Bluffs 712
Des Moines 515
Dubuque 319

Kansas
Topeka 913
Wichita 316

Kentucky
Covington 606
Louisville 502

Louisiana
New Orleans 504
Shreveport 318

Maine
207

Maryland
301
410

Massachusetts
508
Boston 617
Springfield 413

Mexico
52
Mexicali 656
Mexico City 5
Tijuana 66

Michigan
810
Detroit 313
Escanaba 906
Grand Rapids 616
Lansing 517

Minnesota
320
Duluth 218
Minneapolis 612
Rochester 507

Mississippi
601

Missouri
573
Kansas City 816
Springfield 417
St. Louis 314

Montana
406

Nebraska
North Platte 308
Omaha 402

Nevada
702

New Hampshire
603

New Jersey
908
Trenton 609
Newark 201

New Mexico
505

New York
Albany 518
Binghamton 607
Buffalo 716
Hempstead 516
Manhattan 212
New York City 917
Queens 718
Syracuse 315
White Plains 914

North Carolina
910
Charlotte 704
Raleigh 919

North Dakota
701

Ohio
330
Cincinnati 513
Cleveland 216
Columbus 614
Toledo 419

Oklahoma
Oklahoma City 405
Tulsa 918

Oregon
Eugene 541
Portland 503

Pennsylvania
610
Altoona 814
Harrisburg 717
Philadelphia 215
Pittsburgh 412

Puerto Rico
787

Rhode Island
401

South Carolina
Charleston 803
Greenville 864

South Dakota
605

St. Kitts
860

St. Lucia
758

Tennessee
Knoxville 423
Memphis 901
Nashville 615

Texas
Amarillo 806
Austin 512
Dallas 214
Fort Worth 817
Galveston 409
Houston 281/713
San Antonio 210
Sweetwater 915
Tyler 903

Utah
801

Vermont
802

Virgin Islands
809

Virginia
757
Arlington 703
Richmond 804
Roanoke 540

Washington
Olympia 360
Seattle 206
Spokane 509

West Virginia
304

Wisconsin
Eau Claire 715
Madison 608
Milwaukee 414

Wyoming
307

* Area codes as of 1997

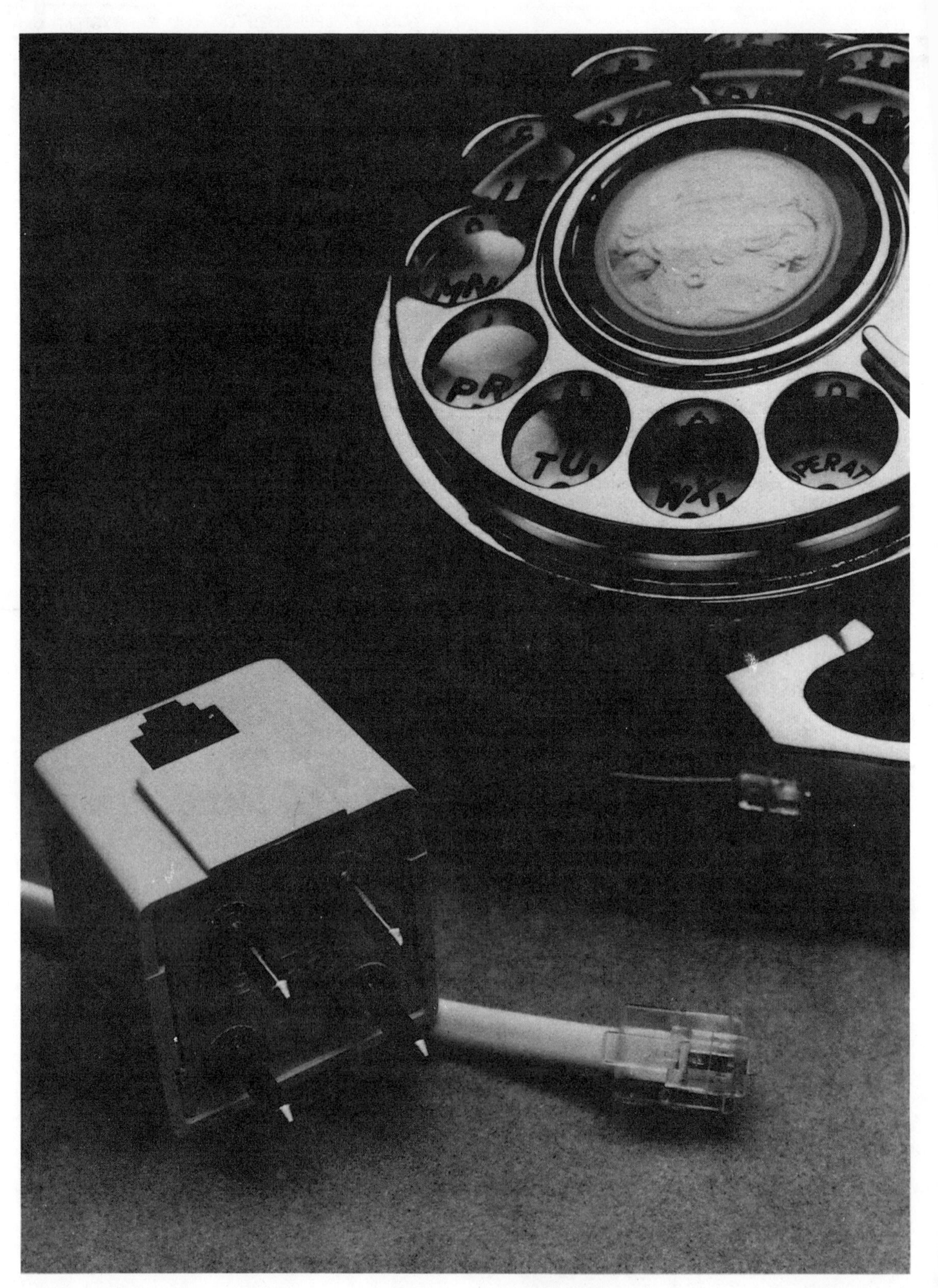

7 ADDING TELEPHONES TO AN OLD STYLE SYSTEM

Chapter Contents *Page*

Case History #1
Simple Techniques 2
Case History #2
Baseboard Installation 3
Case History #3
Attic Distribution 6

If your present house or apartment has an old style telephone system with 4-prong plugs and jacks and directly wired wall telephones and you want to move or add telephones, this chapter, along with what has been presented in Chapter 4, explains how to do it.

After reading this chapter you should:

A. Know some techniques for adding telephones to an apartment or house that has no attic or basement access.

B. Know some techniques for adding telephones to an apartment or house with attic access. The techniques for basement access are covered in Chapter 6.

In each of the chapters — 3, 4, 6 and 7 — the examples chosen try to show different installations. If a particular case that you have in an old style system is not explained, but is covered under the modular system; adapt the material presented on modular systems to the old style system and vise versa.

CASE HISTORY #1 — SIMPLE TECHNIQUES

This case is the same as Case History #1 in the modular system of Chapter 6 except that the system is an old style system. The floor plan is as shown in *Figure 6-1*. The bed in the master bedroom will be moved to the position shown as dotted lines and a desk added in the master bedroom (also shown as dotted lines). An outlet is to be installed at A for a desk telephone, and outlet B is installed to handle the extension telephone by the bed. Outlet P is the existing old style outlet. It is a 4-prong jack outlet like the one shown in *Figure 2-7b* or *Figure 4-7*. The jack is installed in an electrical outlet box.

The installation of outlet A and outlet B is the same as outlined in steps 1 through 8 for Case History #1 of Chapter 6 and illustrated in *Figure 6-3*. The only difference is shown in *Figure 7-1*. Instead of the duplex jack plugging in the outlet directly, a cord adapter is used to convert the 4-prong jack to a modular jack. The duplex jack plugs into the cord adapter and the extension cords for outlets A and B plug into the duplex jack.

Remember to keep the duplex jack unplugged while connecting the wires to the jack terminals for outlets A and B in order to keep from receiving an electrical shock should your telephone ring.

CASE HISTORY #2 — BASEBOARD INSTALLATION

We return to the home used in Chapter 4. The floor plan is shown in *Figure 7-2*. It is the same as *Figure 4-1*, but more detail has been added to show the storage closets and cabinets. Recall that the company telephone line comes in overhead to a protector mounted on the soffit, enters the attic and is terminated in a 42A block. From the 42A block as a junction, one line runs to the counter 4-prong outlet, and thence to the wall telephone at the end of the cabinet above the kitchen counter.

We'll assume that the 4-prong counter outlet has been converted to modular and that the wall telephone has been converted to modular. These installations were described in Chapter 4.

Some rearrangement is in order. You have bumped the table in the hall for the extension telephone a number of times; therefore, you desire to move the extension telephone into the front bedroom and get rid of the 42A block in the hall. You would like to conceal the wiring if you can without climbing up in the attic and making changes. Your solution is to take off the wooden baseboards in the hall and bedroom and run the cable behind the baseboards and up through the wall to a round modular outlet at B. The 42A block will be moved to A inside the closet on the baseboard since presently the cable comes down the closet wall from the attic and through the wall to the 42A block in the hall (*Figure 4-2*.) With the cable hole patched in the hall and the baseboards reinstalled, all the new wiring will be concealed in the hall and bedroom.

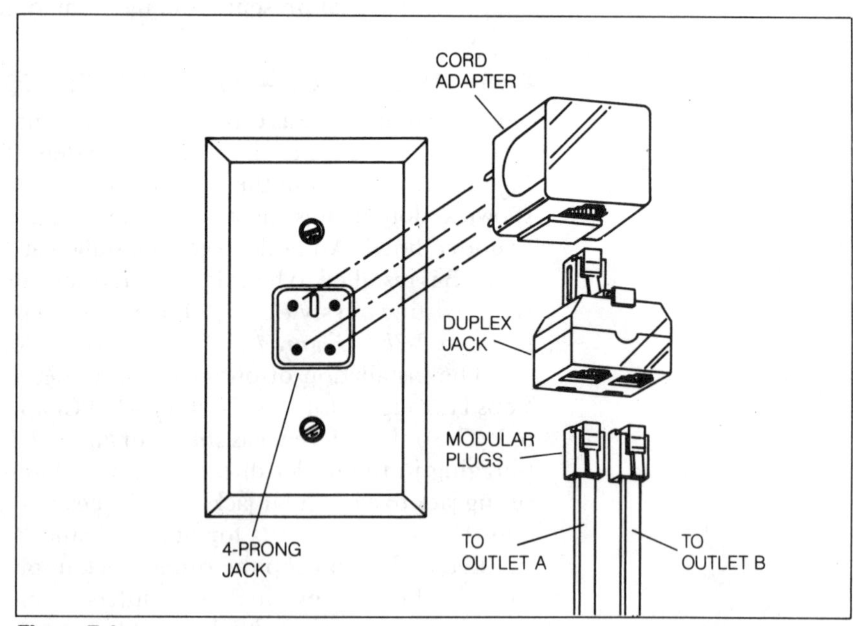

Figure 7-1.
Conversion of 4-Prong Outlet P to Modular to Accept Lines to Outlets A and B

The steps to this installation follow. A layout is shown in *Figure 7-3*. You will have to purchase a round outlet and cable.

Step	*Instruction*
1.	(Remove the telephone handset off hook so there will not be a danger of electrical shock.) With an insulated screwdriver, remove the cover from the 42A block in the hall, and disconnect the wires from the screw terminals. Cut the cable a short length from the end so that all the bare wires will be cut off and will not short to each other.
2.	Go in the closet and pull the cable back through to the inside of the closet. (See *Figure 4-2*.)
3.	Remove the 42A block from the hall baseboard.
4.	Pry the baseboard away from the wall in the hall as required. It may be in one piece down the hall or it may be in sections. Do the same in the front bedroom. Work it carefully a little at a time so you do not split the wood.
5.	Make sure everything is clear of house wiring, plumbing and wall studs below the old hole in the hall wall. Drill a ¼″ hole that will be covered by the baseboard through the wall and through the baseboard in the closet as shown in *Figure 7-3*.

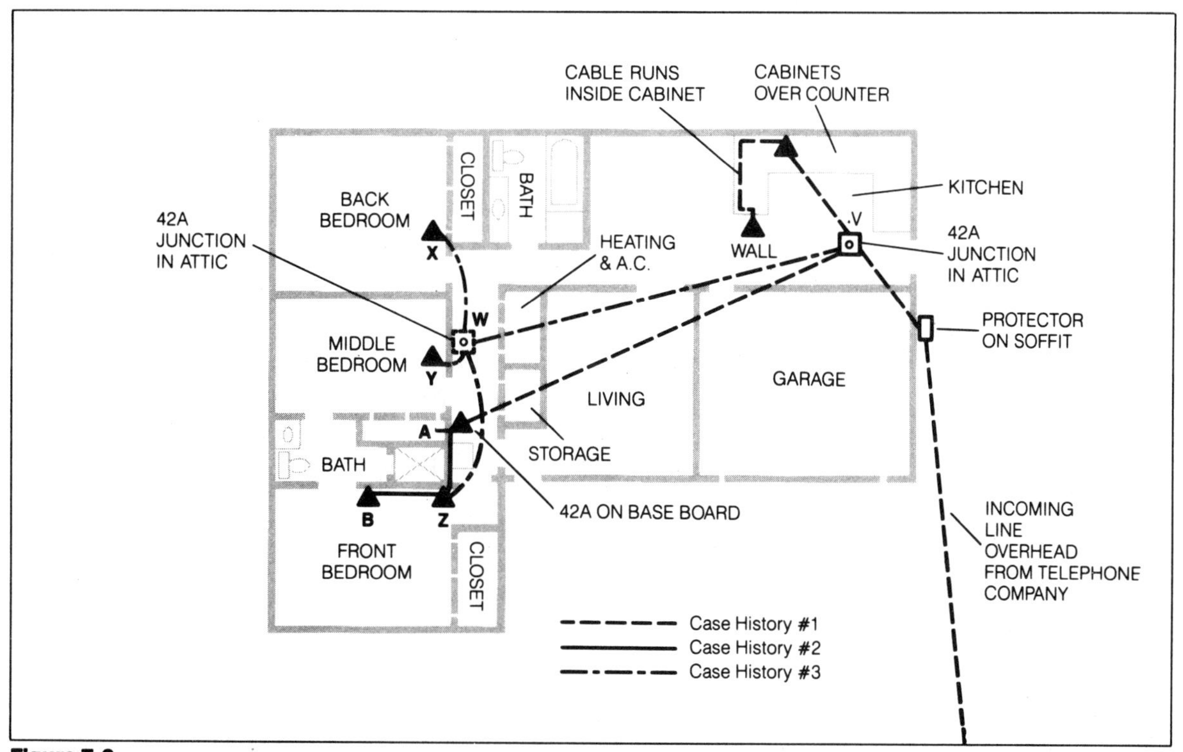

Figure 7-2.
House Floor Plan (Repeat of Figure 4-1)

6. Also, as shown in *Figure 7-3*, drill another ¼″ hole across the corner from the hall to the bedroom in the region that will be covered by the baseboard.
7. Locate the position for the round outlet B in the front bedroom. Make sure it falls between studs. Cut a 1⅜″ hole in the wall at this position. The hole is shown in *Figure 7-3*.
8. In a region to be covered by the baseboard, drill a ¼″ hole into the wall directly below the 1⅜″ hole for the round outlet B.
9. With a knife, chisel or other sharp point, cut a groove in the bedroom and hall walls between the drilled holes deep enough to allow the cable to be run behind the baseboard. *Figure 7-3* shows the cable laid into the grooves.
10. With a nail, string, and a coat hanger or fish tape, fish outlet B and pull the cable from the baseboard hole in the wall to the 1⅜″ outlet hole as shown in *Figure 5-6*. Pull enough cable to leave 6″-8″ sticking out the hole for outlet B (*Figure 7-3*). Attach the mounting ring for the round outlet with plastic anchors and screws.
11. String the cable from outlet B in the front bedroom through the ¼″ hole across the corner and through the ¼″ hole from the hall to the closet. Dress it into the groove as you go.

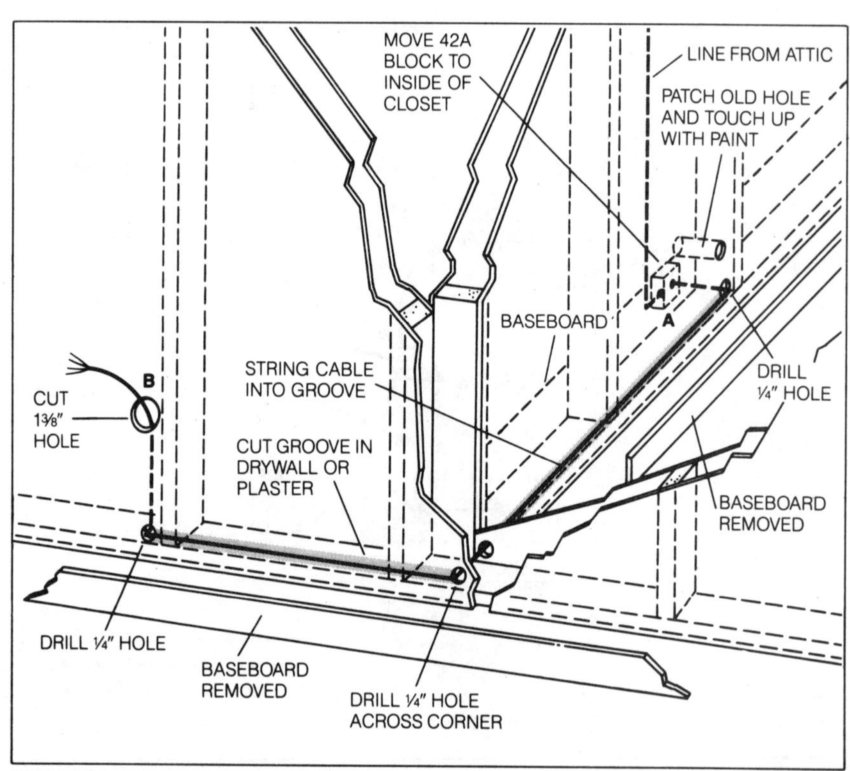

Figure 7-3.
Cable Run Behind Baseboard to Install Outlet B

12. Strip the insulation from the cable and the individual wires and connect the bare wires to the jack body of the round outlet for B — red to red, green to green, yellow to yellow, and black to black (See *Figure 7-6*).
13. Attach the round jack to the mounting ring and screw on the face plate. Outlet B is complete.
14. Go inside the closet and screw the 42A block for A to the baseboard so that the cable coming from the hall can be connected easily to the screw terminals.
15. Strip the insulation from the cables and from each conductor and connect the bare wires to the screw terminals of the 42A block. There are two cables — one from the attic, the other from the hall for outlet B. Match the colors of the wires as you wrap them around the screw terminals. Tighten all screws. Screw down the cover on the 42A block.
16. Test the connections by plugging the extension telephone in outlet B. You should have a dial tone (assuming that the handset that you removed in step 1 has been replaced). If there is no dial tone, check all connections for wires shorting or open connections.
17. Replace the baseboards.

The installation is complete.

CASE HISTORY #3 — ATTIC DISTRIBUTION FOR NEW OUTLETS

On the floor plan of *Figure 7-2* you decide instead of the 42A block and an extension telephone in the hall, you want an outlet in each of the bedrooms. With this plan there could be a telephone in each room, or a telephone could be plugged in if needed. The distribution is to be into the walls from the attic so that all wires are concealed.

Here are the steps for the installation:

Step	*Instruction*
1.	Plan the installation, make a bill of materials and purchase the parts. In this case, you want the wires concealed since the distribution will be from the attic into the walls and down to the outlets X, Y and Z. You decide to disconnect the cable to the 42A block in the hall, pull it back up to the attic, and reconnect it to a 42A block installed in the attic at location W. Location W is a junction point convenient to the cables for X, Y and Z. Round outlets will be installed at X, Y and Z.
2.	(Remove a telephone handset to go off hook so there will be no danger of an electrical shock.) With an insulated screwdriver remove the cover and disconnect the cable wires to the 42A block in the hall. Remove the screws holding the 42A block to the baseboard and save the 42A block and screws for installation in the attic. Clip off the end of the exposed cable so that there will be no bare wires to short out when the cable is pulled back to the attic.

3. Go into the closet behind where the 42A was installed in the hall and pull the cable into the closet (*Figure 4-2*). Now go into the attic and pull the cable into the attic.
4. Find an attic location W that is clear from obstructions and easy to get to and screw the 42A block to a rafter.
5. String the cable from the original 42A junction point at V across the attic to W. Wrap the cable around the rafter so it will not be pulled loose from the 42A block junction points.
6. Go into each bedroom and position the outlets 12″ to 18″ up from the floor in line with any electrical outlets to keep the room uniform. The round outlets are mounted directly to the wall so their position is away from wall studs not next to them. As shown in *Figure 7-4*, there are intercoms in the bedrooms. The holes in the walls cut for mounting the speakers are very useful for fishing the cables through the walls; therefore, a good position for the outlets is between the wall studs directly below the intercoms. Remove the intercom speakers in each bedroom as shown in *Figure 7-4b*.

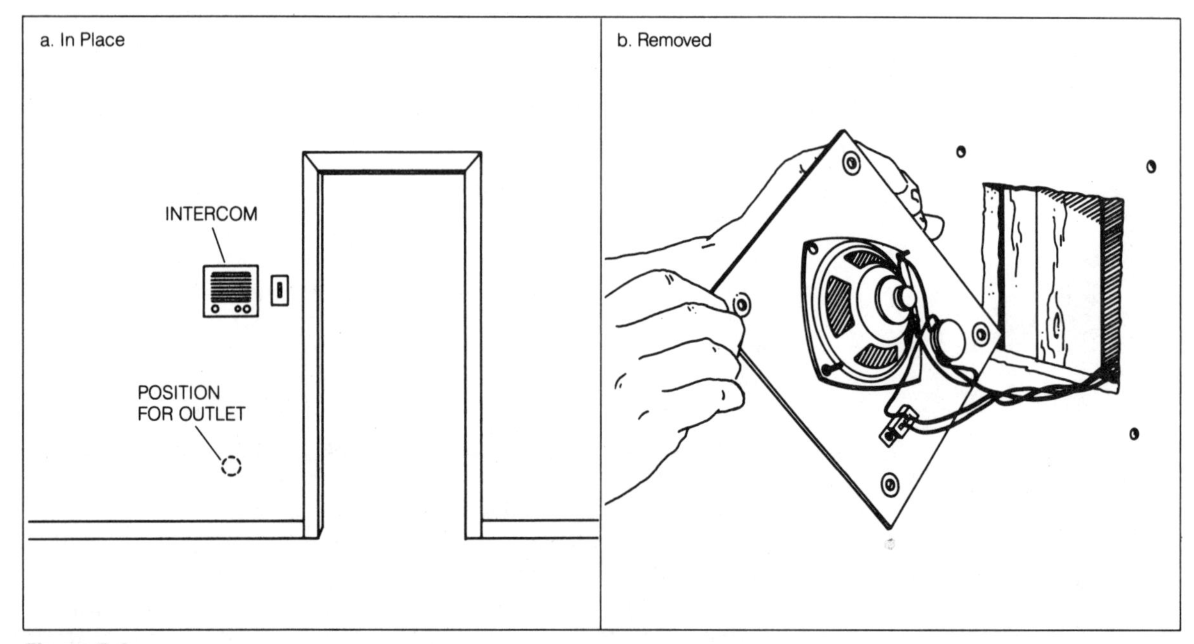

Figure 7-4.
Intercom in Bedrooms

7. Turn to Chapter 5 and follow the instructions in the section "Down From the Attic" to drill the holes in the top plates in the attic. Cut the holes in the wall for the outlets, fish the holes to pull the cables, and pull the cables for the outlets. As a result you should have a cable running down each wall to the outlets X, Y and Z, and extending about 8" out the hole for the outlet.
8. Attach the mounting ring around each outlet hole at X, Y and Z with plastic anchors and screws as shown in *Figure 7-5*. (See also *Figure 5-6*.)
9. Strip 3" of outer insulation from the cable and a ½" from each conductor. Connect the bare wires to the modular jack body as shown in *Figure 7-6* — red to red, green to green, yellow to yellow, and black to black.
10. Screw the bracket that holds the modular jack body into the mounting ring as shown in *Figure 7-7*.
11. Attach the faceplate to finish each outlet installation.
12. Go into the attic, wrap each end of the cable from outlet X, Y and Z around the rafter used to mount the 42A block at junction W. Strip the insulation from the cable ends as in step 9 and connect the bare wires to the screw terminals of the 42A block as shown in *Figure 7-8*. Connect corresponding colored wires — red to red, green to green, yellow to yellow and black to black. Don't tighten the screw terminals as yet. The cable from junction V must still be connected.

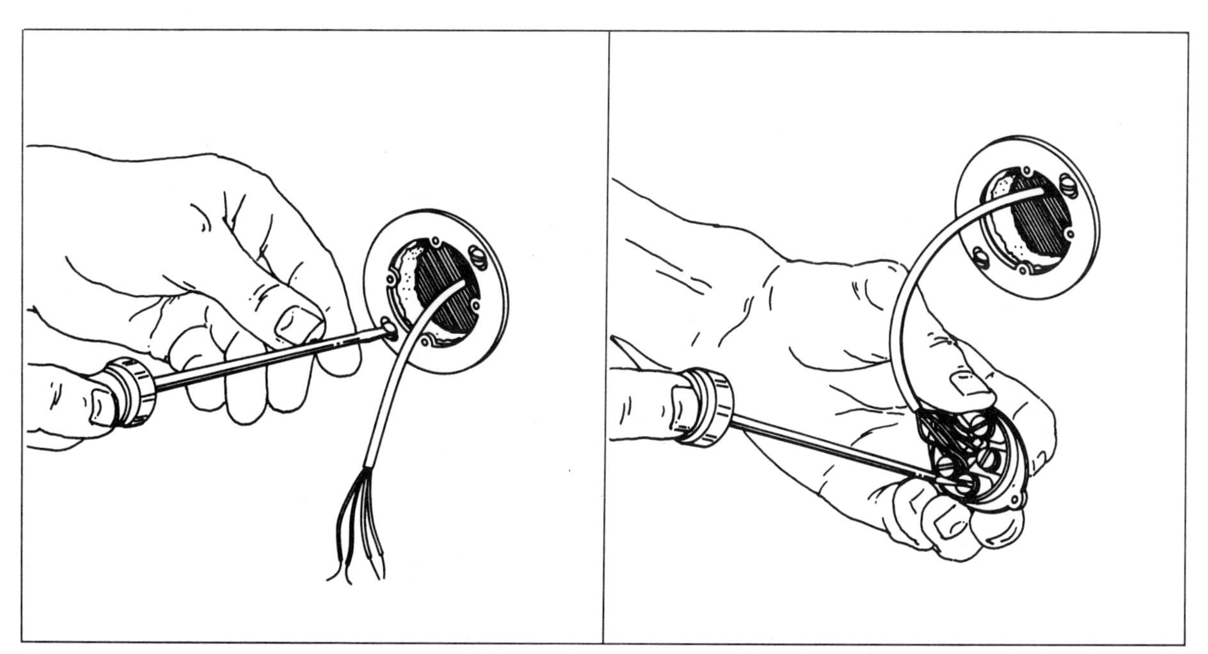

Figure 7-5.
Attaching Mounting Ring

Figure 7-6.
Connecting Cable to Modular Jack Body

13. At junction V, remove the cover of the 42A block and disconnect the cables running to junction W and to the counter outlet. This leaves only the cable from the protector connected to the 42A block at V.
14. Convert the junction V 42A block to modular with a quick-connect modular jack as shown in *Figure 4-4*.
15. Connect the cables that were disconnected from the junction V 42A block to a telephone wiring block by connecting corresponding colored conductors — red to red, green to green, yellow to yellow and black to black. The connection is shown in *Figure 7-9*. Slots in the end of the telephone wiring block need to be cut out to allow clearance for the cables.
16. Connect the cable from junction V to the 42A block at junction W. This is the remaining cable that was left unconnected. Strip the cable insulation and wrap the bare wires around the screw terminals and tighten the screws. Junction W has four cables connecting to it — one from V and one each to X, Y and Z. Keep the correct color combinations. Attach the cover to the 42A block to complete the junction W.

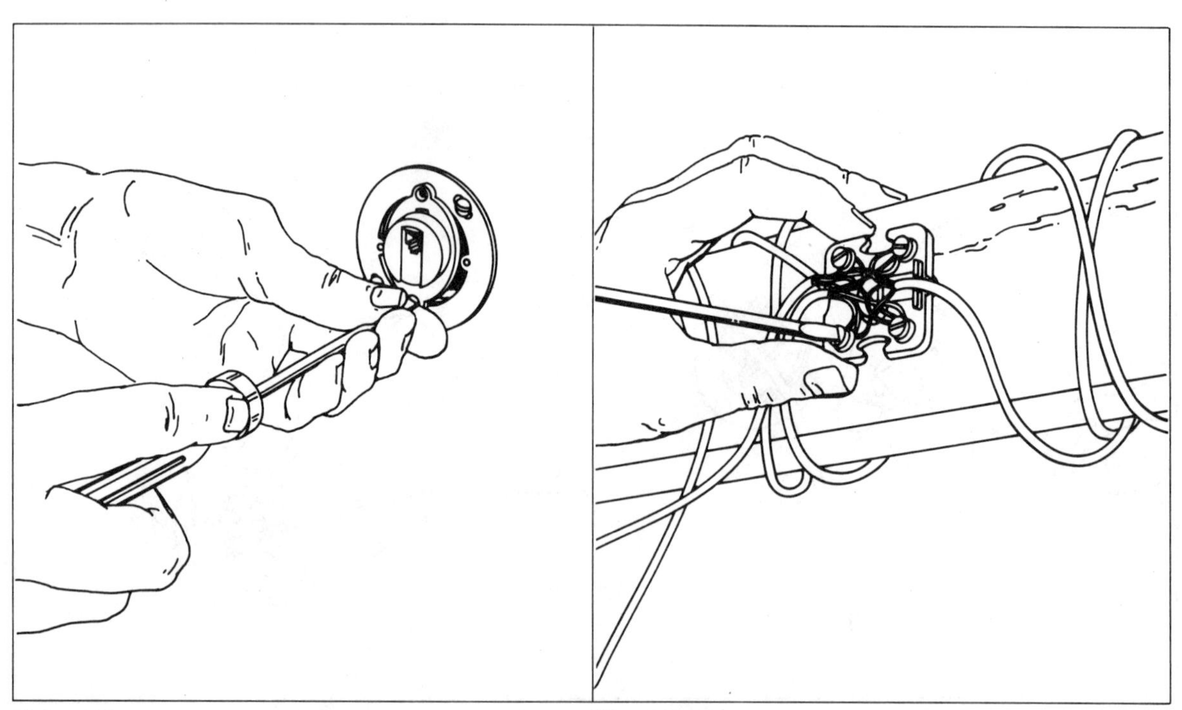

Figure 7-7.
Screwing the Modular Jack to Mounting Ring

Figure 7-8.
Connecting Cables at 42A Junction W

17. All that remains to make all outlets active is to plug the modular plug from the telephone wiring block in the modular jack at junction V. The telephone wiring block is used as a quick-disconnect junction point for all of the replaced or added telephone installations. If there is trouble on the line, all that needs to be done to verify that is in your installed telephones is to disconnect the modular plug of the telephone wiring block from the modular jack. This is a very important connection for maintenance. In fact, this type connection — a wiring block with a modular plug to a network interface that is the initial termination of the telephone line into the building — probably will be the way all local telephone companies will install new telephone service.
18. Test each outlet with an extension telephone to make sure it is active and that a dial tone is obtained when the handset goes off hook. If there is no dial tone, check all connections for shorts or opens.
19. Replace all intercom speakers.
20. Patch the hole in the hall wall where the line came through for the 42A block. Repaint as necessary.

The installation is complete. Extension telephones can be connected in each bedroom and a quick disconnect junction has been installed in the system to make the system easy to maintain.

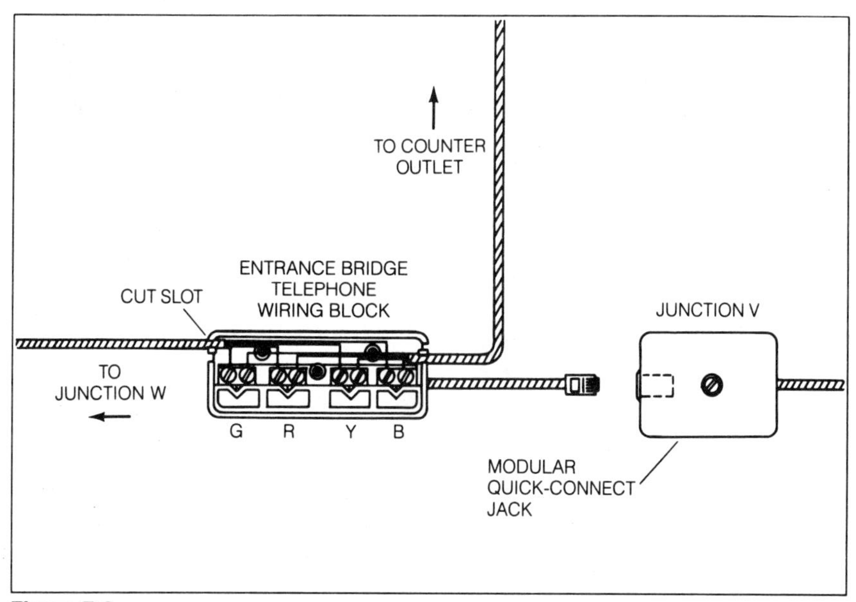

Figure 7-9.
Telephone Wiring Block Connecting to Junction V

PULSE
TONE
1
ABC 2
DEF 3
GHI 4
JKL 5
MNO 6
PRS 7
TUV 8
WXY 9
*
OPER 0
#
HOLD

8 BUSINESS INSTALLATIONS

Chapter Contents *Page*
Single-Line Installations 2
Taps on Multi-Line Service 3
Mult-Line Systems 4
2-Line Systems 5
4-Line Systems 7
Useful Accessories 9

In Chapter 3, a single-party ringing line supplied with a ringing frequency of 20Hz and 30Hz was called a Type-A line. Small business installations that use a type-A single telephone line or multiple type-A lines are discussed in this chapter. Installation instructions for the home and apartment apply equally as well for the single-line or multiple-line telephones for small business. However, there are also the multi-line key-system business installations that are presently in place. Some particular interconnections are possible that add single-line accessories to such multiple-line systems. Any of the devices explained in this chapter are for "standard" systems using Type-A lines.

After reading this chapter you should:

A. Understand what kind of equipment is contained in a typical single-line small business installation.
B. Understand how single-line accessories can be added to a multiple-line business installation.
C. Understand the use of multiple type-A lines in business.

SINGLE-LINE BUSINESS INSTALLATIONS

A typical single-line business installation is shown in *Figure 8-1.* The incoming telephone line is terminated at outlet A, a modular outlet. Outlet B is another modular outlet that is connected to outlet A by any one of three alternatives. Either by an interconnecting cable that runs along the baseboard; or by a cable that runs underneath the floor; or by a cable that runs overhead in the attic or above the ceiling.

The business is arranged so there is a desk isolated in an office and outside the office is a secretary's/receptionist's desk. On the secretary's/receptionist's desk is a telephone that has a feature called automatic dialing. Auto dialers are described in more detail in Chapter 10, but very briefly, automatic dialing is very useful for a secretary because it stores telephone numbers electronically; so that with the push of a button a stored number is dialed. Therefore, frequently used multi-digit business numbers can be dialed by pressing one or two buttons. In addition, to assure that no business calls are missed, a telephone answerer also is installed on the secretary's/receptionist's desk. It can be turned on as required to answer incoming telephone calls.

The auto dialer telephone and the telephone answerer are installed by plugging a duplex jack in modular outlet A to convert the single jack into two jacks. Each of the modular plugs from the answerer and the auto dialer are plugged in the duplex jack to interconnect the secretary's/receptionist's telephone system.

In the office, a telephone amplifier is plugged in outlet B and a desk telephone plugged into the telephone amplifier. The telephone amplifier is particularly useful because the business requires field people to report by telephone to an office staff. With the telephone amplifier, all the staff can hear the report from the field at the same time. The desk telephone handles all of the normal telephone conversations. A buzzer system between the secretary/receptionist and the office (not shown) indicates when the office should answer the telephone because the desk telephone is wired not to ring. There are desk telephones that have built-in intercoms so they are easily substituted and the buzzer system is not needed.

LINE TAPS ON MULTI-LINE SERVICE

4-Line Taps

Many times a business that already has a multi-line telephone system installed would like to add an accessory to the system such as a telephone answerer. A very easy way to accomplish this connection is to use a 4-line tap in the multi-line system.

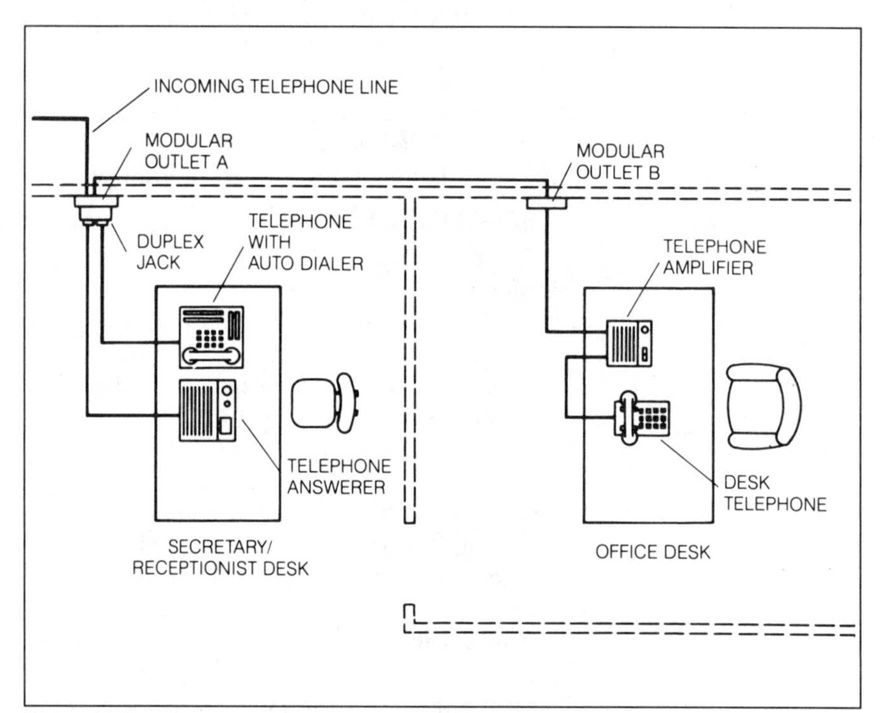

Figure 8-1.
Single-Line Business System

The telephones in the multi-line system are connected with a 50-pin connector as shown in *Figure 8-2.* In order to connect in the line tap, disconnect the male 50-pin connector from the female connector by unscrewing the screws holding the connectors together. Sandwich the line tap between the connectors so the 50-pin male and female connectors match the connectors on the line tap as shown in *Figure 8-3.*

The 4-line tap has four modular jacks that are connected to line 1,2, 3 and 4, respectively, of the multi-line system. As shown in *Figure 8-3,* an auto dialer or a telephone answerer can be plugged into any one of the modular jacks of the 4-line tap. The accessory plugged into the respective line 1, 2, 3 or 4 jack connects to the selected line and can be used as it is used normally on a single-line system.

MULTIPLE-LINE SYSTEMS

One of the ways to increase the capability of the telephone system as a business grows is to install telephone instruments that handle multiple lines. Instead of having only one line to access, telephones of the A type now can access two and even four lines. What makes this particularly attractive is that most of the features, and more, supplied on single-line telephones are found on multiple-line units. With such functionality, multiple-line telephones provide features only supplied previously by key-system or PABX telephones. Two such multiple-line telephones are shown in *Figure 8-4.*

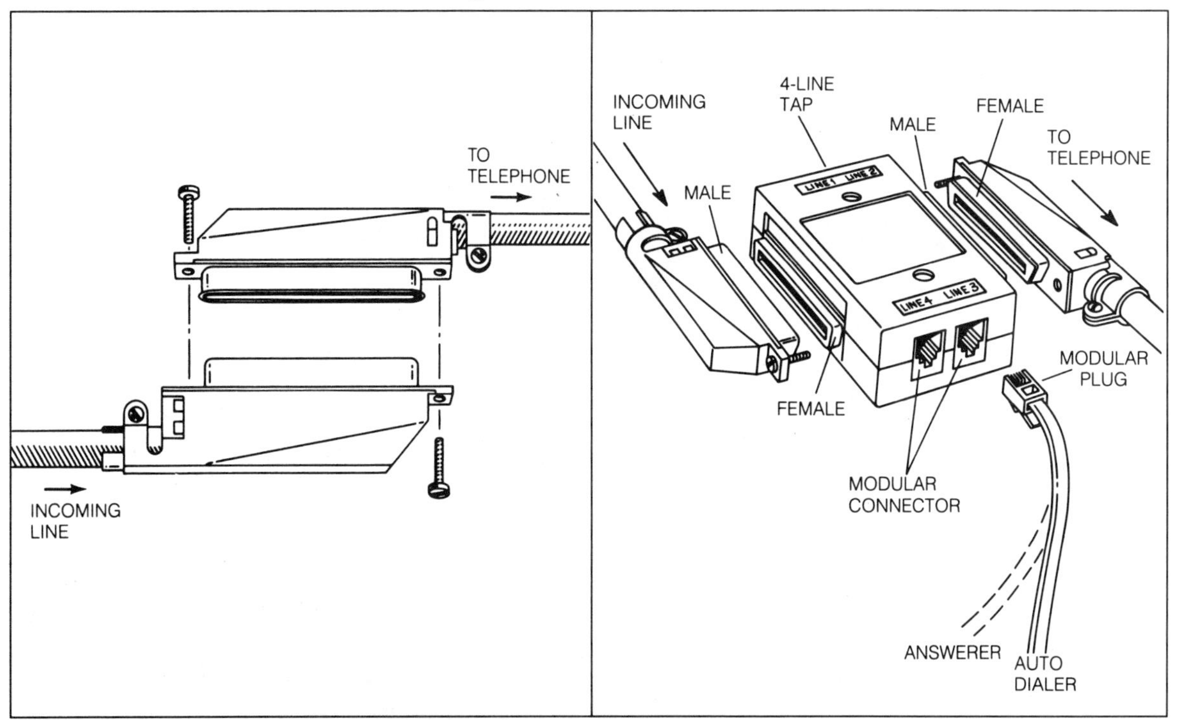

Figure 8-2.
Multiple-Line Connector Separation

Figure 8-3.
4-Line Tap

2-Line Systems

As discussed in Chapter 6, converting from single-line telephones to telephones that access two lines can be as easy as selecting a telephone like the one shown in *Figure 8-4a*, and plugging it into the modular outlet used by the single-line telephones. For example, *Figure 8-5a* shows the same office layout of *Figure 8-1* except another office has been added — actually, three different offices could have been added. The only requirement is that the modular outlets are properly wired as RJ14 outlets rather than RJ11 outlets. If all modular outlets have the four-wire connections of red, green, yellow and black as shown in *Figure 8-6b*, then when line 2 is connected by the telephone company, 2-line service is available to this business.

In *Figure 8-5*, we assume the line 2 connection has been made and all outlets are RJ14. The secretary/receptionist now has a desk telephone at A and a telephone answerer at B, both 2-line units. The desk telephones in the other offices are just like the one at A. They have speakerphones so that the receptionist can talk with each office, and users can have hands-free, on-hook dialing and use. No separate amplifier is required. Each telephone has line status indicators that tell if the line is in use, ringing, or on hold, so that access to a line is easy.

One of the features particularly useful to a business is the conference button providing 3-party conferencing between line 1, line 2 and the calling parties. Memory storage provides calling by name; and an appointment calendar, a call timer, and a date and time display are additional features.

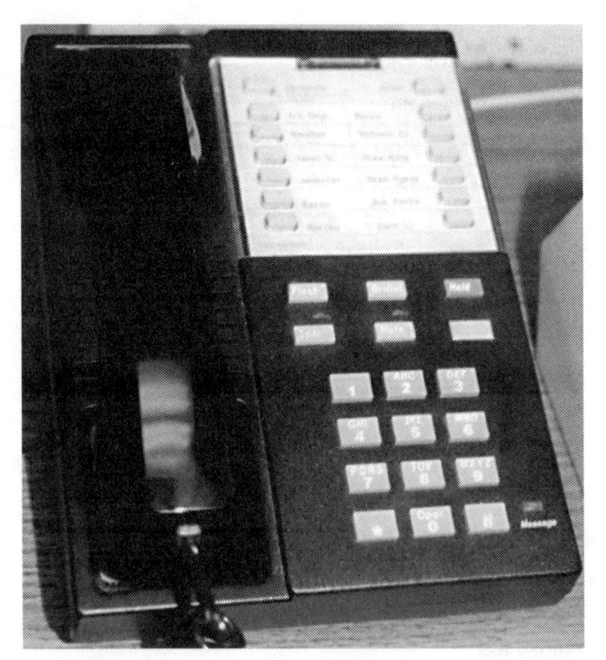

Figure 8-4.
Multiple-Line Telephones for the Office

If the proper connections are not in place, install new cable and new outlets to make all outlets RJ14s.

FM Intercom Network

Some businesses may desire to use a less expensive 2-line system. *Figure 8-5b* shows such a system. A and C are 2-line desk telephones and B is a 2-line answering machine. Communication between offices is now by FM radio frequency signals that allow intercom and paging between stations. Memory dialing, hold button, line status indicators and redial are other features easily provided.

Use of Single-Line Units

As shown in *Figure 8-5c*, businesses should remember that they can still use single-line units on a 2-line system. In A, a controller is installed that allows the single-line telephone at C to be automatically switched between the two lines to select the one not in use. It has ringing/hold indicators. For installation, plug the controller into the RJ14 modular outlet, and plug the single-line telephone into the controller.

In B, an adapter is installed in the other side of the duplex RJ14 jack to allow the use of a single-line telephone answerer on both lines. The adapter automatically switches the answerer to the ringing line.

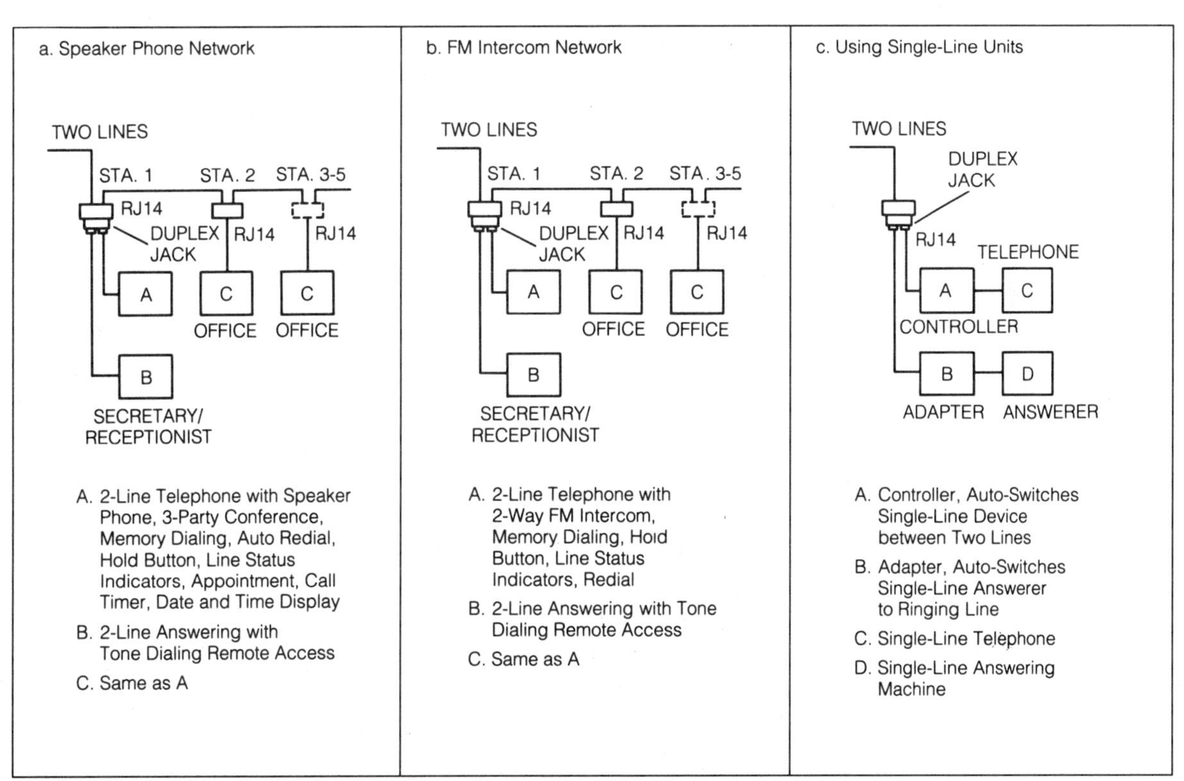

Figure 8-5.
2-Line Business Systems

4-Line Systems

Businesses that have expanded sufficiently to require access to four telephone lines now have available 4-line telephones that are Centrex/PABX compatible. Features such as line status indication, hold, intercom, paging, on-hook dialing and "hands-free" calls, flash button for call waiting, privacy and "do-not-disturb" buttons are now available and controllable for the four lines at the desk telephone. And with no additional wiring than normally used for telephone line interconnections discussed in this book.

For example, *Figure 8-6* illustrates the system line connections required for single-line, 2-line and 4-line systems. Single-line telephone systems *(Figure 8-6a)* really only require two wires — red and green — And RJ11 modular connections, but we have stresses several times that usually the four wires — red, green, yellow and black — and the RJ14 modular connections are already in place for 2-line systems *(Figure 8-6b).*

Expansion to a 4-line system requires cabling for two 2-line systems as shown in *Figure 8-6c.* Therefore, if one set of RJ14 modular connections like *Figure 8-6b* are already in place, the only requirement for a 4-line system is to install another set of RJ14 modular outlets. The RJ14 cable and outlet installations use the techniques described in Chapters 5, 6 and 7. A typical business installation with five telephones is shown in *Figure 8-7.*

Typical Installation

In *Figure 8-7,* all stations have 4-line telephones. If full-feature, top-of-the line telephones are selected, most or all of the features listed in *Figure 8-7* can be provided. Less-expensive models do not provide speakerphone or automatic dialing, but the important intercom and paging functions are maintained. An amplifier for speakerphone operation and an autodialer can be added as accessories at stations where they are desired.

Station 1 is considered a secretarial/receptionist station and has an answering machine that handles two lines. The answering machine (presently connected to line 1 and 2 through the duplex jack shown) is connected to the lines most likely to receive calls in off-hours. If a shift were required to line 3 and 4, the duplex jack would be moved to the adjacent RJ14 outlet at station 1.

Station 2 through 5 may have top-of-the line telephones or lesser-expensive models may be used. With such an arrangement, the secretary knows the status of all lines, can put any of them on hold, can speak with any station or page all stations, can have on-hook dialing and hands-free calling, and can transfer calls to required locations. Only five stations are shown, but the system is expandable — up to 10 without overlapping intercom numbers; up to 12 with overlapping intercom numbers.

As shown in *Figure 8-7,* an AC outlet for each telephone is required. Some telephones operate directly from 120 VAC; others operate using an adapter to a lower voltage. Adapters are shown in *Figure 8-7.* A 9V battery also is required to maintain memory storage and to power the ringer if 120 VAC power fails.

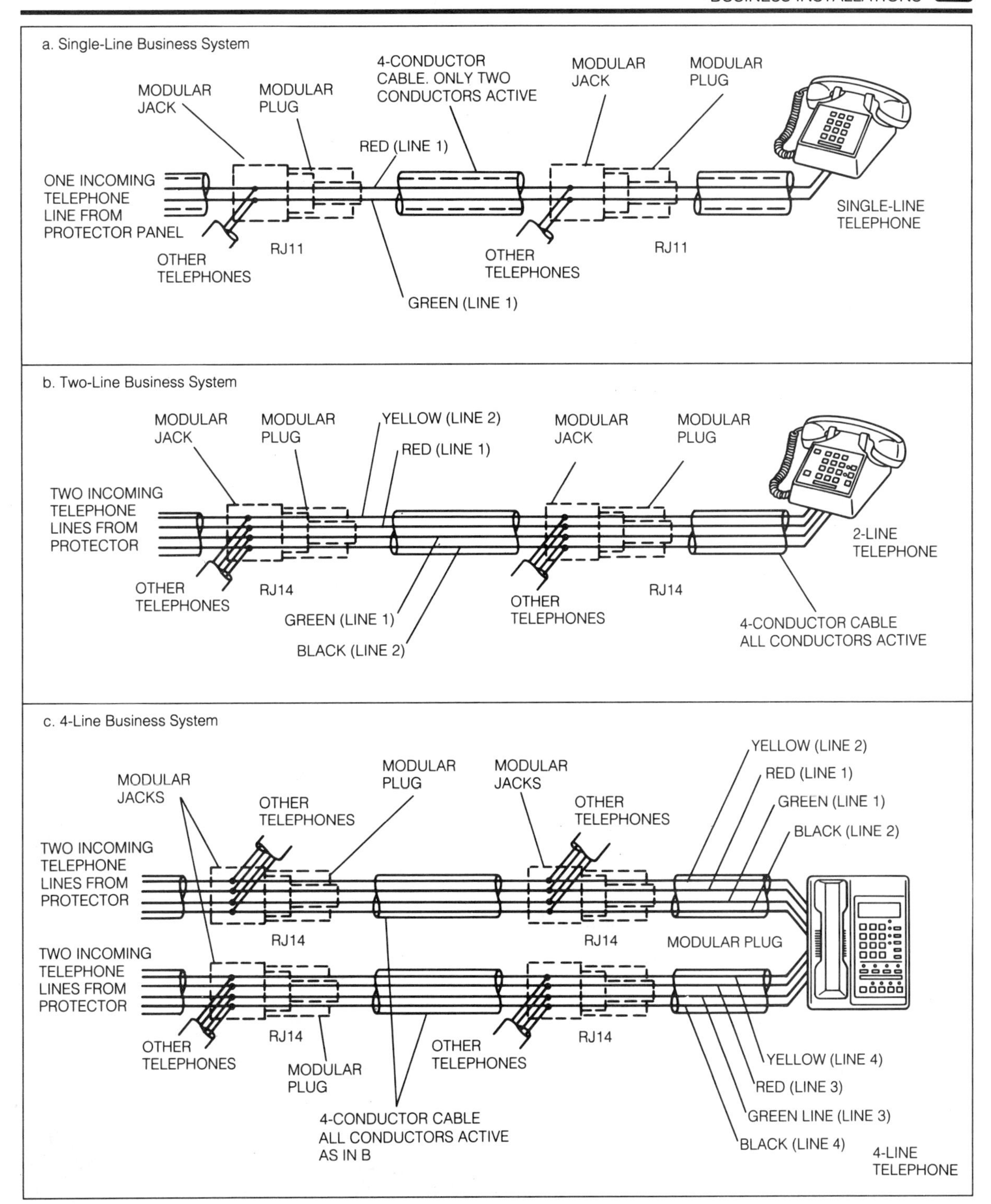

Figure 8-6.
Distribution of Business Systems with One, Two and Four Lines

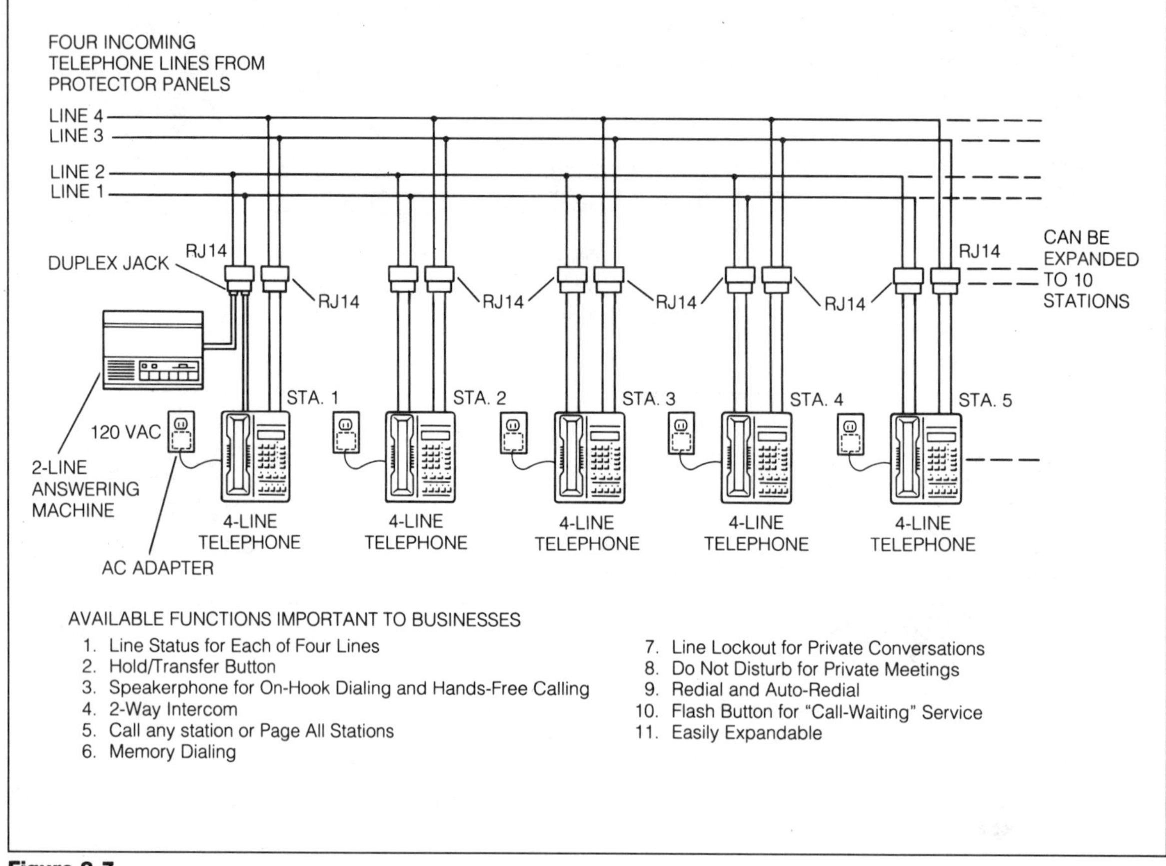

Figure 8-7.
Business System with Four Lines

With the RJ14 modular outlets in place, all that is required for an operating system is to plug in the RJ14 plugs from the telephones and plug in the adapters or a line cord into 120 VAC. It is assumed the telephone company has been called to connect the lines at the central office.

USEFUL ACCESSORIES

Hands-Free Telephone Headset

An accessory that is quite useful in business for hands-free, fatigue-free conversations is a telephone headset like the one shown in *Figure 8-8.* Receptionists, in particular, have a real need for such a unit. It has an adjustable mike boom, can be worn on either ear, can be switched from handset to headset operation, and adjustable volume from the amplifier provides clear and concise communications. The control unit with amplifier can be attached to an existing telephone connected to an existing modular outlet.

Recording Controllers

In business there is a need many times to record telephone calls completely and accurately. Special accessories are available that couple the telephone line to a cassette recorder and start and stop the recorder automatically. They are called recording controllers and use a cassette recorder that has a remote mike jack.

Basic Types of Recording Controllers

1. A single telephone recording controller records only from the telephone to which it is connected.
2. A multi-extension controller provides the capability to record conversations from any extension on the telephone line.

How to Connect Recording Controllers

Installing a recording controller is quite simple as shown in *Figure 8-9.* A duplex jack is used in the modular telephone outlet to connect the input to the recorder and the existing telephone to the telephone line. Two plugs from the recording controller go to an audio tape cassette. One plug is the audio to the cassette, and the other plugs in the remote mike jack to turn the cassette recorder on and off.

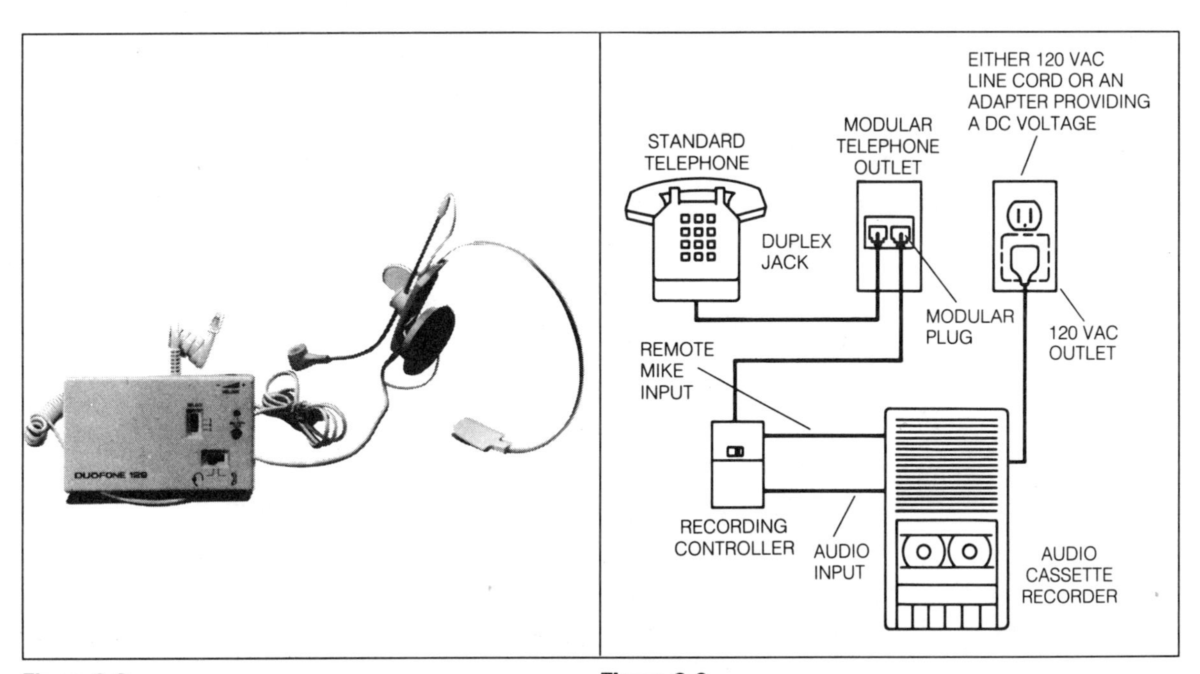

Figure 8-8.
Telephone Headset
Courtesy of Radio Shack

Figure 8-9.
Telephone Recording Interconnections

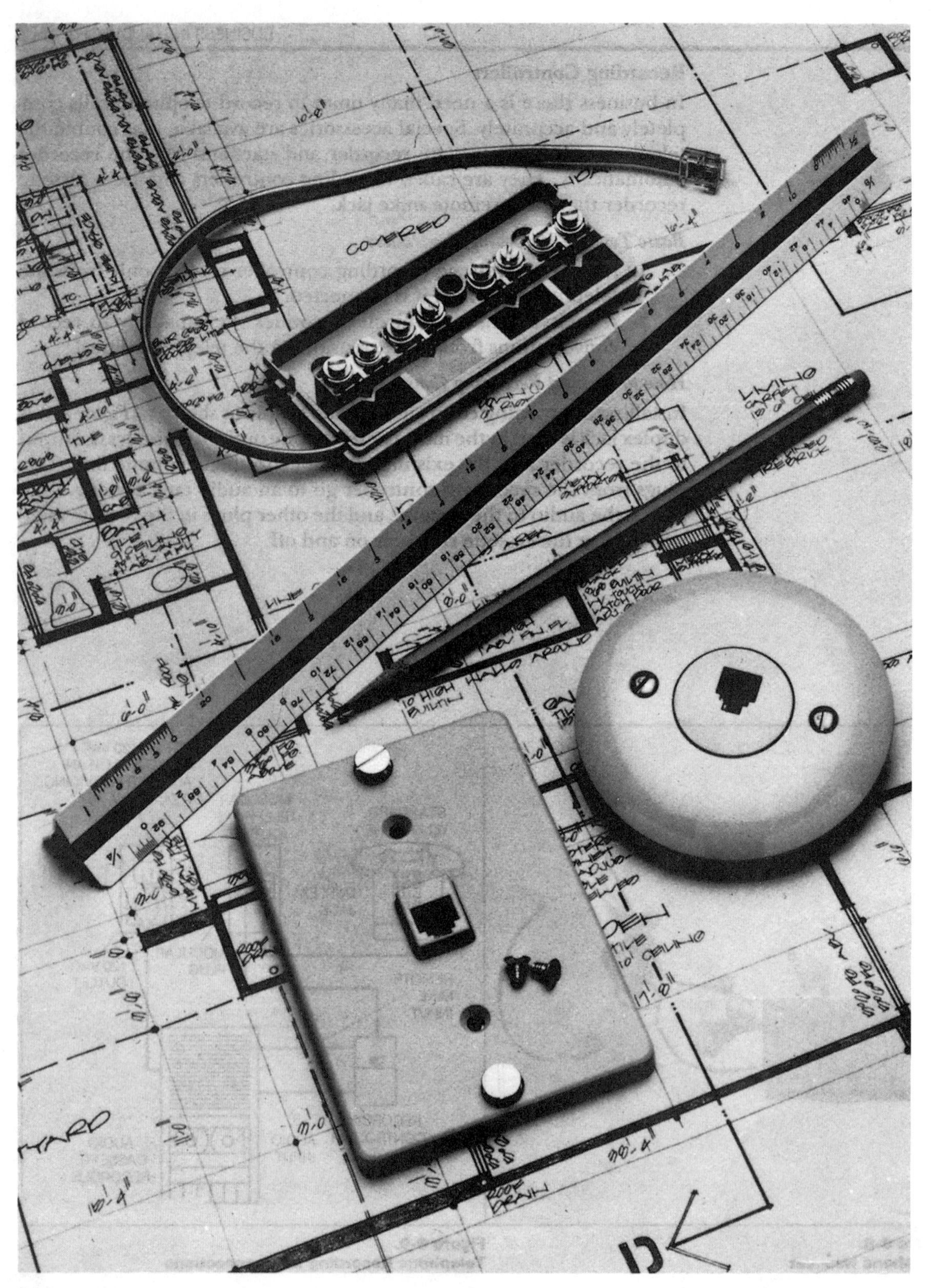

COVERED
LIVING
BATH
DEN
TILE
10' CEILING
10 HIGH BUILT IN
ROOF DRAIN

9 PREWIRING INSTALLATION

Chapter Contents	*Page*
Installation Standards or Codes	*2*
National Codes	*2*
Local Codes	*3*
Floor Plans & Layouts	*3*
Bill of Materials	*7*
Rough-in Boxes and Wiring	*12*
Final Installation	*16*

If you are building a new home, townhouse or condominium, there is an opportunity for you to save some money by prewiring your telephone outlets. In addition, by planning the installation yourself, you can have your telephone outlets exactly where you want them rather than having them determined by a contractor or installer.

After reading this chapter you should:

A. Be aware that there are standards and codes for wiring.
B. Be able to make a layout of a prewiring telephone installation on a floor plan.
C. Prepare a bill of materials.
D. Be able to rough in boxes and wiring in a structure under construction.
E. Complete a final installation of jacks and wall plates for modular outlets.

INSTALLATION STANDARDS & CODES

National Codes

Here are some important points from the National Electric Code (Article 800 Communication Circuits) that should help you install a safe, reliable telephone system in your home or business.

- Do not locate or place telephone cables in pipes, conduits, boxes, or compartments with other electrical wiring.
- Do not place telephone cables near bare electrical wiring, antennas, transformers, lighting rods, pipes (water, steam, gas), heating ducts or an exhaust flue.
- Protect telephone wires from abrasion with electrical tape or tubing when they may contact rough surfaces such as gratings, grill work or masonry.
- Do not use telephone wires or cables to support anything, such as lighting fixtures or decorative items.
- Do not run telephone cables between structures where they could be exposed to lightning. Either relocate or when the telephone company makes the service connection have them install a protector at each location where a cable enters the building.

- Do not install telephones or outlets in damp locations or anywhere where the location would allow a user to use a telephone while in a bathtub, shower, swimming pool, or sauna. If a short were to develop in a telephone, a shock could result.
- Do not locate telephone cables where they may be subjected to physical strain, pinching or cutting from such things as door jams or window sills.
- Never splice or twist telephone wires together directly. A splice can corrode and may cause interference on your line. Use wire junctions, connector blocks, or connectors as outlets for making connections.
- Keep cable and wire runs as short as possible and practical. For best results, it is recommended that no more than 5 telephones be connected to any one telephone service line (see Chapter 2).
- Before drilling holes through floors, walls, studs or plates, check carefully to avoid drilling into pipes, electrical wiring or ducting.

Local Codes and Information

Contact your local City, Township, or County Building Inspector and request information or copies of all local codes which apply to telephone wiring or communication circuits. Review this information to assure your installation will comply with the local codes.

Call or visit your local telephone company business office and request information they may have available on prewiring and any special requirements they may have for user-wired installations. Find out whether or not your telephone service will be underground (buried) or overhead (aerial drop from a utility pole).

FLOOR PLANS AND LAYOUTS

Your prewiring installation will be aided a great deal by making a layout on a floor plan. Planning your layout will be simplified if you can obtain a copy of the floor plan of your home, apartment or small business from the builder or architect showing the electrical wiring. The electrical symbols on the floor plan will be:

⦶ Wall Outlet
—+ Wall Switch
-○- Ceiling Light
-○) Wall Light

The architect also may have noted his desired location of telephone outlets using the symbol ▲ and usually noting which symbols are for wall phones.

Basic Structures

The basic structures that will be considered are:

A. Single Story Residence

1. With underneath crawl space or basement — this will allow access to the walls from under the floor.

2. On a slab construction — only access to the walls will be from the attic.

B. Multiple Story Residence

1. With underneath access — each level may be considered as a different plan with access from the lower floor level.

C. Mobile Home

1. Some preconstructed mobile homes come prewired for telephones; in most cases, there will be little opportunity for prewiring. However, there will be opportunity to install the telephones initially. Follow the instructions for adding telephones to a modular system in Chapter 6 and the instructions for running cables in Chapter 5. If telephone outlets exist and they need to be changed, follow the instructions in Chapter 3, 4 or 5.

Layout

The layout of your telephone system on the floor plan is important. It is a guide that will save you time making the installation; provide the means for the listing of your bill of materials; and provide you a *lasting* record of the telephone wiring for future reference.

Before starting your telephone layout on the floor plan, determine:

A. Is there going to be just one single line or are there going to be multiple lines for your children, your mother, father, or your business?

B. Where are future add-ons such as cordless phones, wall phones or extension phones going to be placed? Now is the time to run cables for those anticipated additions.

C. Is there going to be a security system? If there is, whether or not it is a separate hard-wired system or used with the telephone system, now is the time to run the cables. Running the cables is quite similar to running telephone cables.

D. How about a computer system? Maybe there are plans now or maybe a system is planned for the future. If so, running cables for such a system might be in order.

The first thing to do is study the floor plan and determine the location on the outside of the structure where the electrical service from the power company will come to the electric meter. Then go outside the structure and verify the location. This will be the location of the service entrance where the telephone company will mount the lightning protector and connect their service line. The telephone company chose this location to have a common place for entry into the structure and because good grounding is available.

The floor plan of a typical single story residence with the planned telephone layout is shown in *Figure 9-1.* The electrical wiring has been omitted from this figure because all the symbols would cause a great deal of clutter on this size drawing.

Figure 9-2 contains the plans and telephone layout for a two-story residence. A profile is included showing how the wiring flows from where it enters the structure through the structure to provide the outlets on all three levels.

You may have noted that in both of the layouts of *Figure 9-1* and *Figure 9-2* a complete loop is formed that starts from the service entrance where the telephone company will connect the protector and returns to that same point. It has an A leg and a B leg. Solid 24 gauge wire cable should be more than ample for most telephone installations, but as added protection, by running the complete loop with A and B legs, you will have two parallel circuits of 24 gauge wire to carry the electrical load of your total telephones. In addition such a loop connection provides an alternate cable path in the event that a cable is severed or cut at some later date.

Some telephone companies recommend the use of 22 gauge solid wire telephone cable because of the electrical current demand of electromechanical ringer telephones. The installation of up to five (5) telephones could create an electrical overload on cable with wire size smaller than 22 gauge.

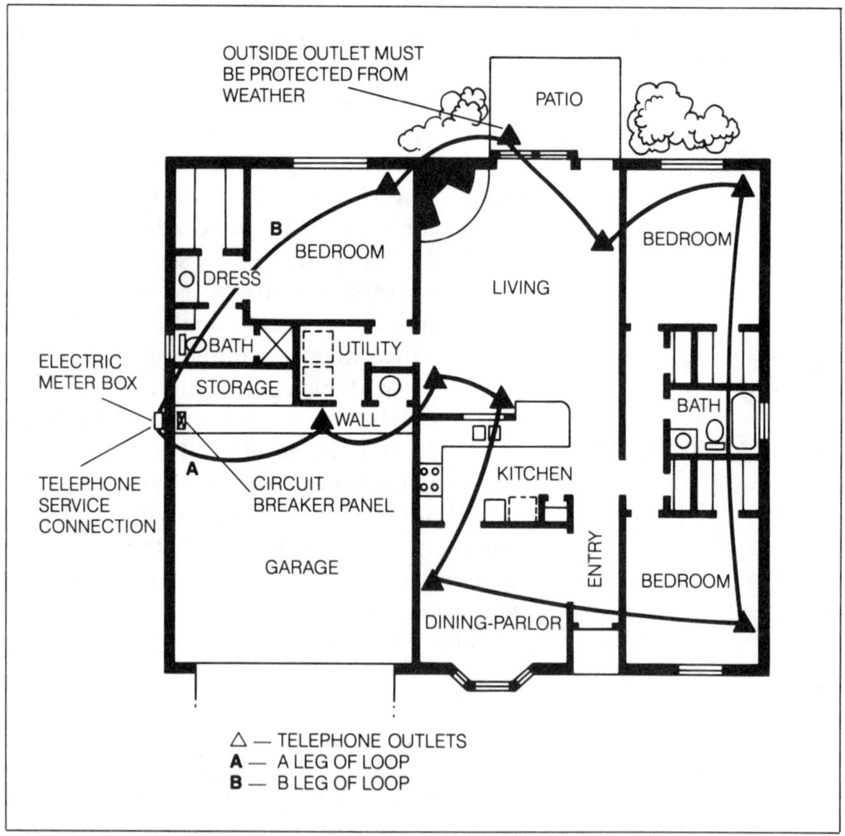

Figure 9-1.
Single Story Residence Plan

Telephone Locations

The location of telephone outlets, though not critical, needs to be in convenient places that provide flexibility for room arrangements and future expansion. As a result, some thought should go into the placement of telephone outlets. The following check list should be useful in making the placements.

Check List

A. Location of telephone service into the structure — Is this in a convenient location for repair and maintenance?

B. Location of telephone outlets

1. Which has highest priority? Usually kitchen, master bedroom, a location central to children's room, den or living room are first priority.
2. Which has next priority? Remaining bedrooms, a bathroom, a game room, the garage, the pool area, a patio, maybe an outside outlet would be possibilities.

C. Bedroom outlets should be placed near a corner that is common to a bed position that might be located on one wall and in the future changed to another wall.

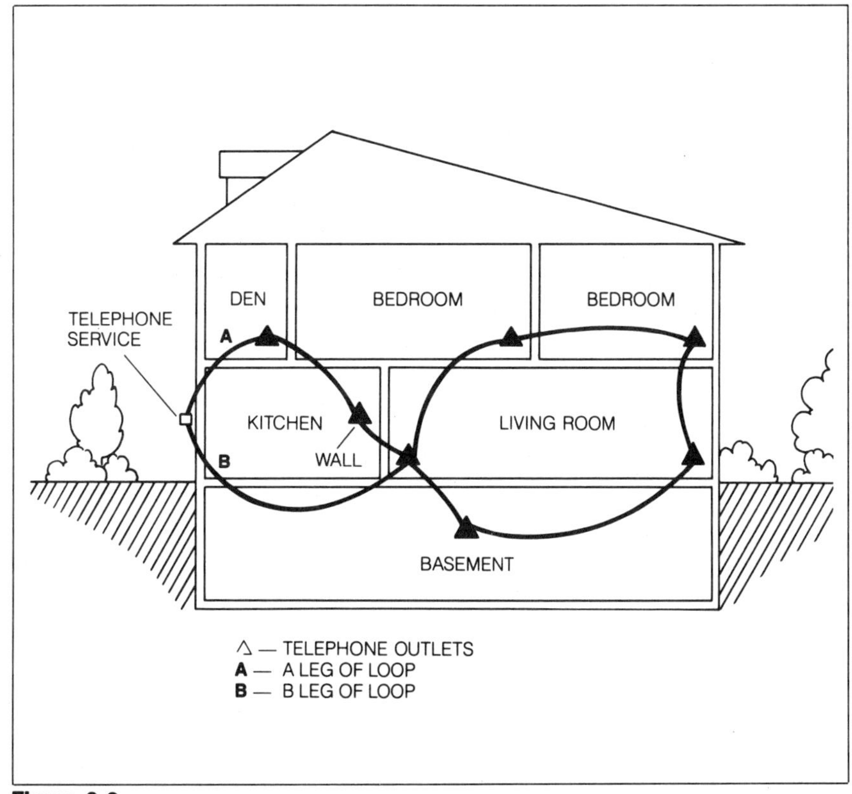

Figure 9-2.
Two Story Residence Plan

D. Kitchen wall outlets above a counter should be placed high enough to allow the cord to clear the counter. With the variety of telephones that can be used either as a wall or a desk/counter telephone, the kitchen outlet might be placed in a location that serves both mountings.
E. Cordless or full-feature telephones will require 120 VAC power or an adapter. Locate modular outlets that service such telephones close to electrical outlets.
F. Flasher modules, dial-light transformers and other similar telephone accessories may also require 120 VAC; therefore, electrical outlets may be required for them.

BILL OF MATERIALS

The bill of material is created so that the items necessary for rough-in can be purchased and so that an estimate of the total cost of the installation can be obtained including the telephones to be purchased. *Figure 9-3* is designed to aid you in developing your bill of material. Since all the present day telephone equipment is equipped with modular connections, only modular outlets are listed for new prewiring construction. You must know the scale for your floor plan layout drawing in order to estimate the approximate telephone cable lengths required. Obtain the parts from your local electronic parts or telephone supply store.

Cable Lengths

Use the following as guidelines for various installations. To figure the vertical cable drop from the attic to an outlet, use a length of eight feet (8′) and from underneath the floor to an outlet use two feet (2′). For a wall phone outlet, use a length of four feet (4′) from the attic and six feet (6′) from underneath the floor. Most residential walls are eight feet (8′) from floor to ceiling; however, be aware that vaulted ceilings and industrial/commercial wall heights may be ten (10′) or twelve (12′) feet high.

When scaling out the telephone cable lengths, you may use the chart to tabulate the lengths. Start at the service entrance to the first telephone outlet; add the length to each outlet to the drop from the attic or rise from the floor underneath; then record the determined length on the chart. Go from outlet to outlet using the same procedure. Don't forget to include the length between outlets. After all lengths are recorded, total the lengths. Add 30% to the total length. This compensates for the extra wire that will be necessary to conform to the various codes and provides adequate length for wiring procedures that will be described later.

Bill of Material Items

A description of each item on your Bill of Materials table of *Figure 9-3* will follow. This will help you decide where to place them on your plan.

Entrance Bridge Telephone Wiring Block (Figure 9-4)

Use this wiring block as the distribution point, terminal block or junction point if the telephone company installs an inside modular jack (usually called a modular interface) as the initial connection from the protector. This wiring block has a 12" cable with a modular plug for direct connection to the modular jack provided by the telephone company. It becomes a "quick disconnect" for your installation. Local telephone companies probably will require this type of installation in the future.

42A Connector Block (Figure 2-6c)

Primarily this is a terminal block used for initial service termination or as a junction point. It has screw terminals and may be used to connect bare wires stripped of insulation or wires that have spade lugs on the ends. Use this connector to splice wires when making the cable runs or place it in a location that will be easy to get to for future expansion or for troubleshooting.

ITEM	Kitchen	Family Room	Living Room	Den	Master BR	BR #2	BR #3	BR #4	Bath MBR	Bath #2	Garage	Patio	Total
Entrance Bridge Wiring Block													
42A Connector Block													
Modular Outlet Rect.													
Modular Wall Outlet													
Wiring Wall Box*													
Weather Proof Cover*													
Modular Outlet Round													
Mounting Bracket**													
Outdoor Ringer													
Flasher Module													
Cable 100′													
Cable 50′													
Wall Telephone													
Desk Telephone													
Combination Wall/Desk													
Cordless Telephone													
Fashion Telephone													

*Local Hardware or Builders Supply.
**Local Electrical Supply (Western Electric 63B, Suttle 63A or equal.)

Figure 9-3.
Bill of Materials

As an example, in *Figure 9-1* with attic distribution, locate a 42A block in leg A of the loop in the attic prior to the drop into the garage wall and one in loop B prior to the drop in the bedroom wall. Place this connector block in a location that will be easy to reach after the structure is completed, such as close to a disappearing stairway or an attic access opening. For the plan shown in *Figure 9-2*, it is suggested that a 42A block be located in leg B of the loop in the basement prior to going up through the floor to the outlet in the living room wall which will also serve the outlet on the kitchen wall. A 42A block would not be necessary in leg A because it would not be accessible after the structure is finished. The cable itself will be accessible in the den outlet box.

Round Modulator Outlet (Figure 2-10a)

This wall outlet is preferred by some because it is round and stylish. Note that the round modular outlet will not cover the box circuit opening nor will the bracket match the screw mounting holes of the wall box. A special mounting strap (*Figure 9-5*) is often available in electronics stores. It should be mounted to a stud during rough-in if an in-wall mounting plate is desired. If the mounting strap is not used, the round modular outlet must be mounted directly in the wall during the final installation of outlets. Installation will require a 1$^{3/8}$" hole in the wall for the body of the outlet, with the outlet mounting ring fastened with screws and plastic screw anchors in drywall and with screws in plaster/lath or wood paneling.

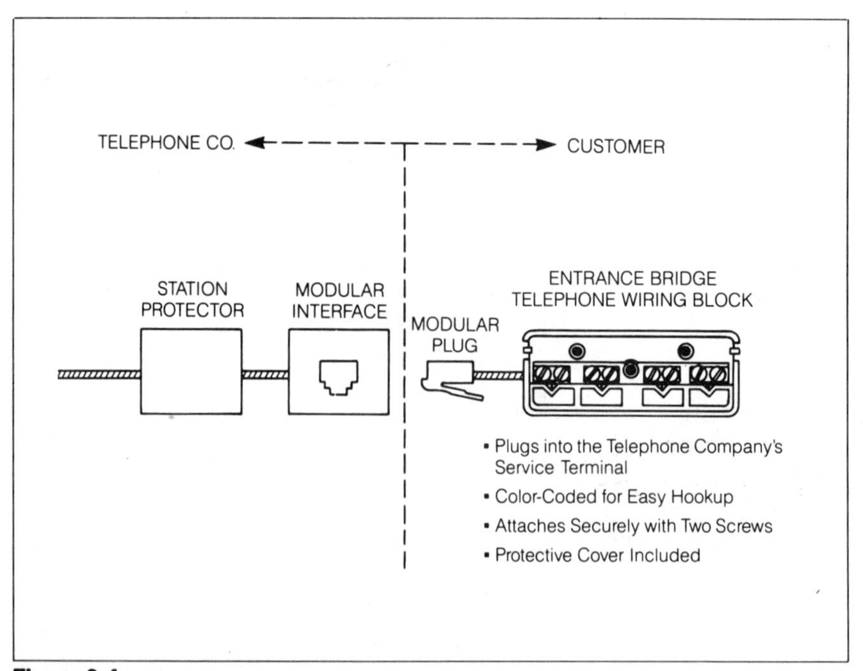

Figure 9-4.
Telephone Wiring Block

Rectangular Modular Outlet (Figure 2-10a)

This outlet fits into a standard type electrical wall box *(Figure 9-5)* and may also serve as a terminal block for wire splicing of cable that runs to the next outlet.

Wall Telephone Plate (Figure 2-9)

This wall outlet is to be used where a telephone is to be mounted on a wall. This outlet mounts into a standard type electrical wall box *(Figure 9-5)*. This outlet may also serve as a terminal block for wire splicing of cable runs to the next outlet.

Standard Type Electrical Wall Box (Figure 9-5)

This is the type of electrical box also used for electrical light switches and electrical outlets. You may purchase these at your local hardware or builder supply. The metal type ($2 - $4) or the PVC plastic type ($1 - $2) are the most common.

Weatherproof Outlet Cover (Figure 9-6)

This is similar to the standard switch or single outlet weatherproof unit used for outside electrical outlets except it has a single modular outlet. The cover may be purchased from a electrical supply house, but make sure that it will adequately cover the telephone outlet. A separate modular outlet must be installed, and the weatherproof outlet cover will be used in place of the plastic faceplate.

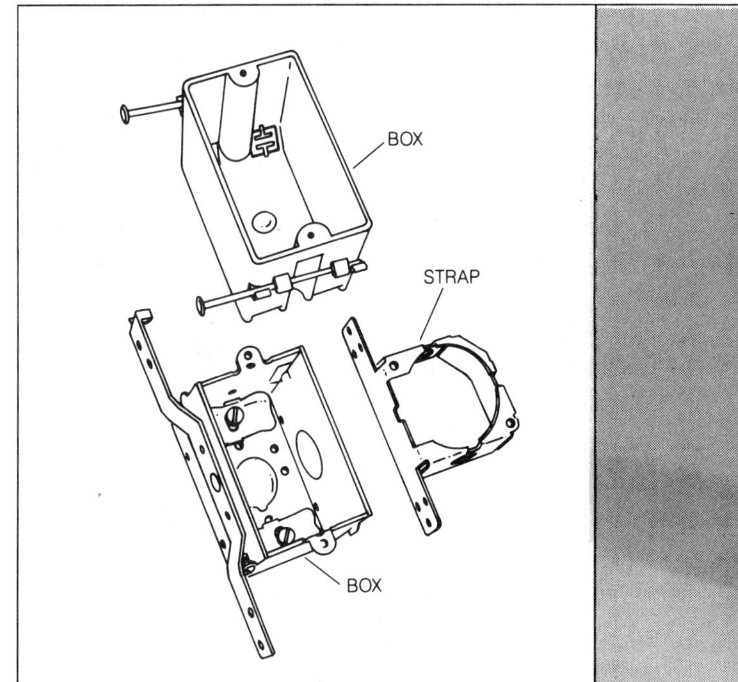

Figure 9-5.
Standard Type Electrical Wall Box

Figure 9-6.
Weatherproof Outlet Cover
Courtesy of Radio Shack

Outdoor Phone Ringer (Figure 9-7)

This may be used outdoors, such as on the patio, in the garage, or workshop where a loud telephone ringer is desired. This outdoor ringer requires a short modular cord from a modular outlet to the ringer's modular jack. Locate the weather resistance ringer under the roof overhang or similar location where it will not get direct rain or water exposure. Installation details are shown in Chapter 10.

Telephone Flasher Module (Figure 9-8)

This unique unit has three modes of operation when there is ringing voltage on the line indicating an incoming call. It flashes a built-in strobe light or it sounds a buzzer that's louder than a normal telephone ring, or it does both. The unit can be mounted in a convenient visible location near a telephone modular outlet and a 120 VAC outlet. The unit has a 6.5 ft. power line cord that is plugged into 120 VAC for power and a 5.5 ft. modular cord with modular plug that is plugged into the telephone outlet. Additional details on flasher installation are discussed in Chapter 10.

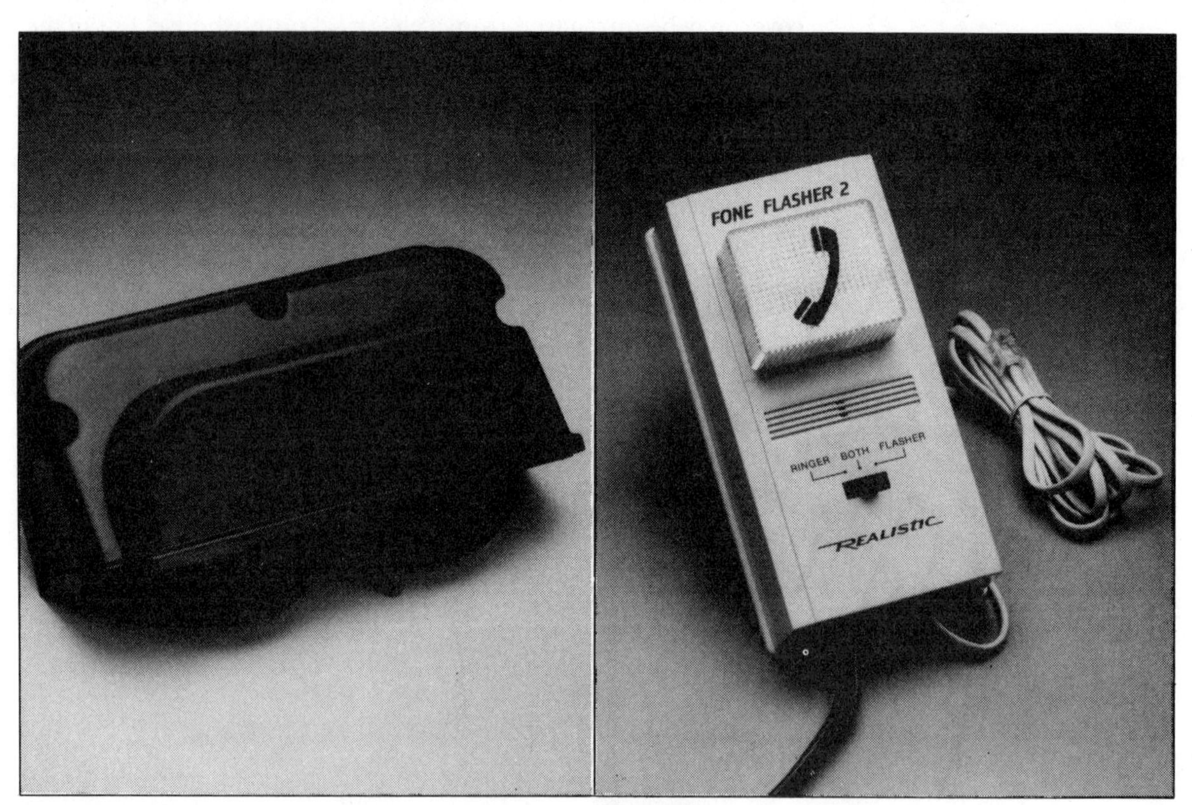

Figure 9-7.
Outdoor Ringer
Courtesy of Radio Shack

Figure 9-8.
Flasher Module
Courtesy of Radio Shack

4-Conductor Phone Cable (Figure 9-9)

Standard telephone cable used for residential wiring is 4-conductor (2 pair) 24-gauge solid wire cable. It usually comes in 50′ or 100′ rolls. It will be the cable to use for telephone cable runs from the telephone company service to each outlet designated on your plan or layout.

Telephones

The telephones listed in the bill of material table *(Figure 9-3)* are not itemized but are listed by the basic functional type because of the large selection of different styles and types. As you make your selection, write in the model number and stock number for future reference.

ROUGH-IN BOXES AND WIRING

Contact the electrical subcontractor or electrican who is doing the electrical wiring and coordinate the proper time to rough in the telephone lines and boxes. Usually the convenient time is right after the electrican has completed the electrical rough-in by setting the main power panel, outlet, switch boxes and completed electrical cable runs.

Start by marking the locations for all boxes for rectangular outlets *(Figure 2-10a)*, all boxes for wall mounted telephones using the wall plate *(Figure 2-9)* and the placement of any round modular outlets *(Figure 2-10a)*.

Figure 9-9.
4-Conductor Solid 24-Gauge Telephone Cable

Boxes for Outlets in Walls

For a rectangular outlet, attach an electrical box *(Figure 9-5)* to the stud (vertical 2 × 4) at the same distance from the floor as the electrican has set the electrical boxes. This distance will vary by locale and by local building code but usually will be 12″ to 18″ from the rough floor. The telephone outlet box may be located on a common stud adjacent to an electrical outlet box *(Figure 9-13)* or on a stud by itself. Mount the box such that it protrudes forward from the stud the thickness of the wall covering to be installed later. Use as a guide the same dimension as the electrican used for mounting the electrical outlet and switch boxes.

Boxes for Telephone Wall Plates

Most wall telephones are 50-52″ above the floor, but you may choose the height to best suit you. Mount the outlet box the proper distance from the floor and again with the proper distance forward of the stud so it will be flush with the wall covering that is installed later. The thickness of common drywall is ½″.

Round Outlets

If the round outlets for modular connectors are to be mounted to the wall without a strap, the location is just noted so a hole can be left in the wall covering and the cable pulled through. Therefore, drive a nail or staple in the stud where round modular outlets are to be located *(Figure 9-11).* This is a temporary marking and will only be used until the cable at this location is pulled through the wall covering. If a mounting strap is to be used, mount it to a stud just like an outlet box.

Telephone Service Entrance

Next, locate the electrical service entrance. For underground telephone service, locate the telephone service cable hole 12″ below the actual or estimated electrical meter box position and 6″ either side of the electrical service entrance conduit *(Figure 9-10b)*. For overhead telephone service *(Figure 9-10a)*, locate the telephone service cable hole 12″ above the actual or estimated electrical meter box position and 6″ either side of a vertical line from the meter box. *Caution:* The location of the electrical wiring in the wall from the meter box to the power panel may well dictate the telephone cable location. If the above locations were to interfere with or come in contact with any wires, move the entry location to another place that meets the codes discussed previously. Drill a ⅜″ hole in the structure siding for the telephone cable if the siding is already present. Caulk the cable hole to protect from insects and moisture.

Laying The Cable Runs

Start the telephone cable from the hole in the wall for the telephone service entrance. Push the cable through the hole and leave 18″ of cable on the outside wall *(Figure 9-10c)*. Route the cable vertically (up or down) along a stud to the plate. The plate on the bottom of the wall will be one 2″ × 4″ stud and on the wall top a double 2″ × 4″ stud as shown in *Figure 9-13* and *Figure 9-12* respectively. Drill a ⅜″ hole in the plate as necessary, run the telephone cable along the joist (ceiling or floor depending on your type of installation) working toward the next outlet. Locate the cable such that it will not be disturbed (cut, torn, or broken) by workman later as they finish the structure. Route the cable along the joist wherever possible, not across them to keep the cable protected. Where the cable must cross joists, you may want to drill holes in joists to make the crossing run. Drill the necessary holes in the plates to allow a drop from the attic or a rise from underneath to the next outlet.

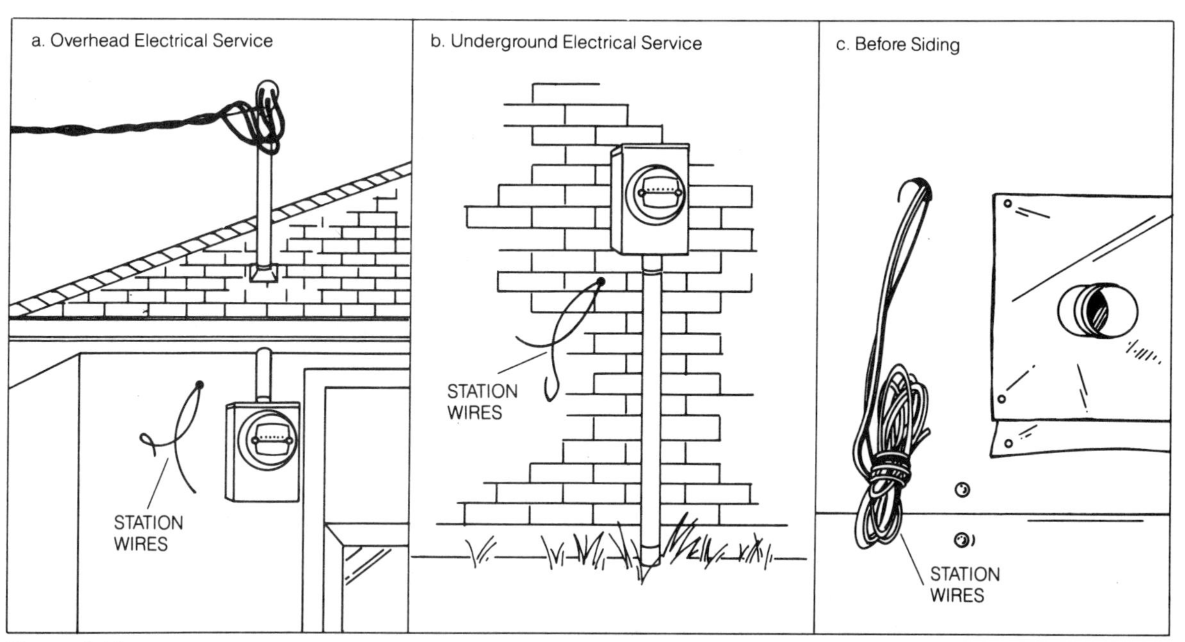

Figure 9-10.
Telephone Service Connection

A typical attic distribution run is shown in *Figure 9-12*. A drop to the basement is shown in *Figure 9-13*. Use romax electrical cable staples to secure the phone cable to studs and joists locating one every 3 to 4 feet of cable. Drive the staple in just far enough to hold the cable in place, but leave ample space between the staple and the stud to allow the cable to be pulled out for later rework or troubleshooting. A typical installation is shown in *Figure 9-12*. From the box run the cable back to the plate hole and continue on to the next outlet in the same manner stapling as you go. Take the excess slack out of the cable as you complete your run.

In the event you have not used a round outlet mounting strap, you will not have an outlet box for the loop but instead a rough in staple as shown in *Figure 9-11*. In order to do the final installation of the round outlet after the wall is covered, you will have to carefully measure and record measurements in order to find the cable. Once the cable is located, a 1⅜″ hole is drilled in the finished wall to accept the outlet body, terminate the cable and mount the outlet. Another way is to tag the cable with instructions that tell the carpenter to bring the cable thru a hole in the wall at the location of the staple or nail while attaching the sheet rock or wall paneling.

If any cable splices are required, use the 42A block in the cable run, screwing down the block screws over bare wires to make reliable connections. Otherwise, lay in the cable lengths to allow for cable splices in the outlet boxes at the screw or slot terminals of the modular connector bodies of the outlets.

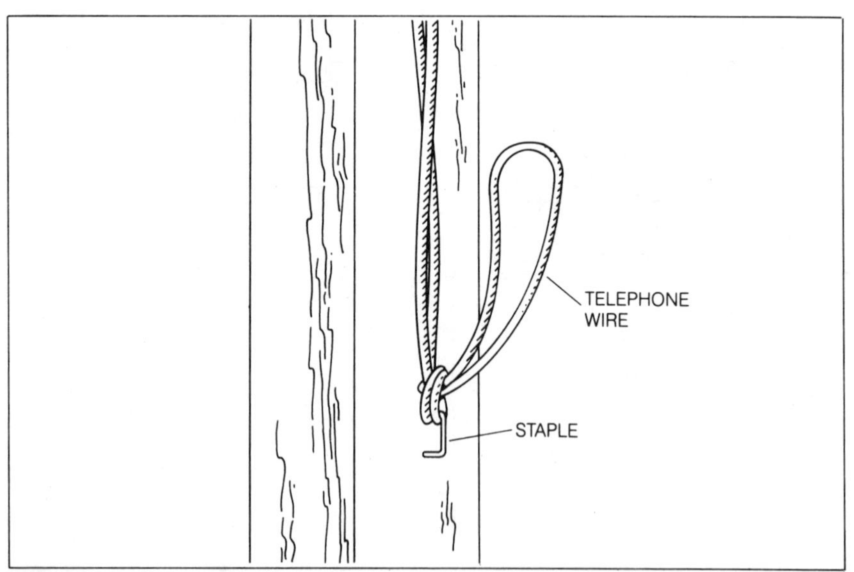

Figure 9-11.
Cable Loop Around Staple in Stud

Complete Circuit (Continuity) Test

Once you have completed the entire wiring route and completed the home run back to the service entrance, or wiring block, label the legs of the loop A & B respectively. It is suggested you conduct a continuity check on the completed rough-in to make sure you have no broken wires or cables that would interrupt a complete circuit to your telephone. If you have any cable that is not looped at an outlet box but has two ends in the box, strip the cable insulation back and separate each wire. Strip the insulation from each conductor and expose ¼" of bare wire. Then twist the bare wires (for test only) of like colored wires together, red to red, green to green, etc. At the service entrance if the wires are still in a loop, cut the cable and strip cable insulation and each conductor insulation to expose ¼" of bare wire from each conductor. You now should have four (one for each color conductor) loop circuits with the ends open at the service entrance. Use a continuity meter, or ohm meter, and check across the ends of like colored wires (black to black and so on) of the legs A & B of the loop cable at the service entrance. You should have continuity. Refer to troubleshooting Chapter 11 if you do not have continuity in the loop circuits.

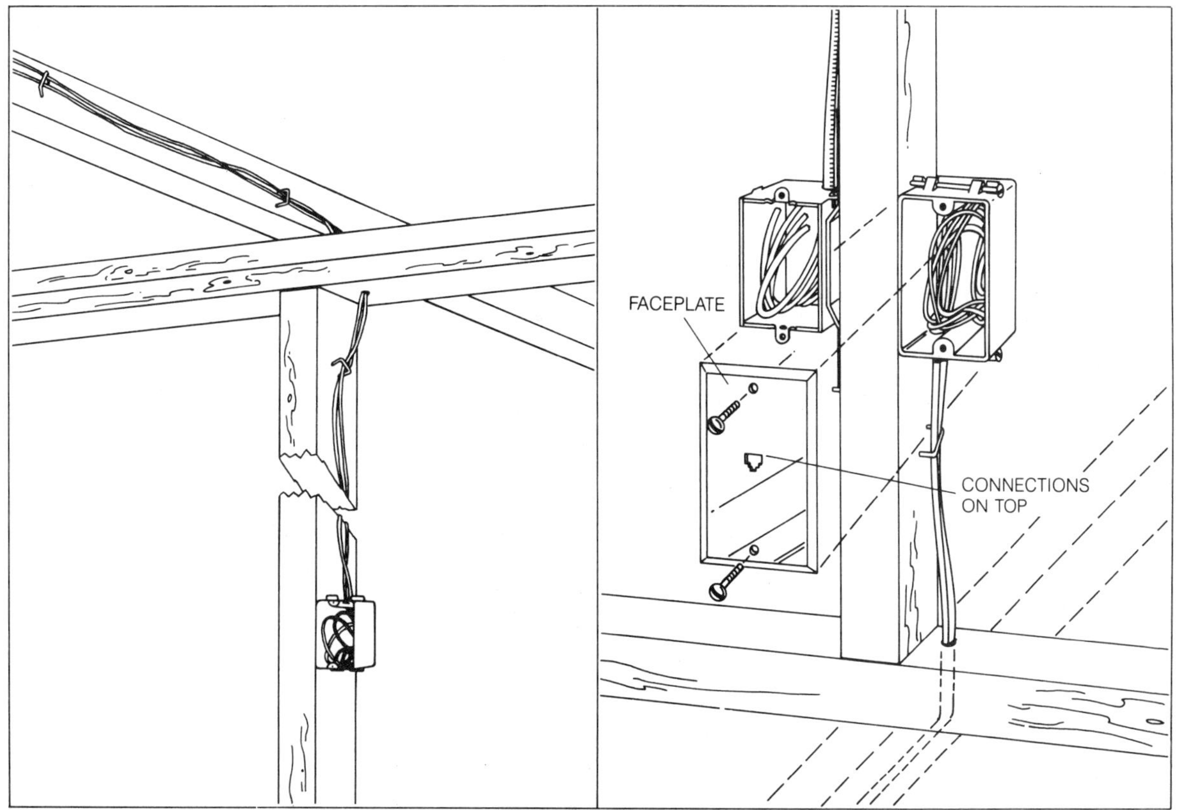

Figure 9-12.
Cable Running to Attic and Along Joist and Down Another Wall

Figure 9-13.
Cable Running Up from Floor

Rough-In Cleanup

Once you are satisfied the rough-in is electrically complete, go to each outlet box and coil up the wire in the box to keep it out of the workman's way when they complete the wall covering. When the structure is completed and the electrican completes his installation of the outlets and switches, you may complete your finish connections of outlets and wall plates. Again, coordinate this final installation with the electrician or builder.

FINAL INSTALLATION

You must complete the interior installation of all outlets and finish mounting faceplates and wall telephone plates before the telephone company installs the protector and connects the service to the central office. Start on leg A of the loop of your plan and work through the loop circuits to the first outlet. Install any mounting plate, the outlet body, a faceplate and proceed on to the next until all outlets are connected and installed in place. An example of installing a faceplate is shown in *Figure 9-13*.

Follow the instructions packed with each outlet regarding wire connections and mounting. Some helpful hints are:

A. Adjust the cable loop at the outlet location to have 8″-12″ of cable exposed from the box or coming out the wall. Push cable back into the wall as required.

B. You may cut the wires at each outlet to break the loop or you may strip insulation from the cable and each conductor to expose bare wire and wrap the bare wire around the screw terminals of the outlet body.

C. To strip outer cable insulation (sheath) remove 3″ of the outer insulation from the cable. Be careful not to cut through the conductor insulation when removing the cable insulation.

D. Some outlet connections have metal slots that pinch the conductors to make connection. Do not remove the conductor insulation (red, black, yellow, green, or see Chapter 5 for other cable wire combinations) if the outlet has slot connecting type terminations. Push the wire covered with insulation into the slot, the metal slots will break through the insulation and pinch the wire to make connection.

E. Screw terminals require stripping the insulation to expose bare wires. Remove the conductor insulation ¼″-½″ with a wire stripper. Be careful not to cut or nick the solid copper conductor. Wrap the bare wire around the screw terminal in a clockwise direction, the same direction that you tighten the screw.

F. Another way to connect the screw terminals and not cut the loop is to strip the sheath insulation from the cable for 3″ to 5″ and separate out the conductors. Without breaking the wire, strip the insulation from the wire for about 1″. Wrap the bare wire around the screw terminal in a clockwise direction and tighten the screw. Connect red to red, green to green, yellow to yellow, black to black, etc.

G. Mount modular jacks so the notch is at the bottom and the brass rail contacts are at the top (See *Figure 9-13*).
If the round modular outlet will not be mounted in the mounting strap but directly into the wall, remove the wall plate and outlet body from the metal mounting plate. Mount the plate to the wall with plastic anchors and screws prior to cutting the 1⅜″ opening in the wall in the center of the bracket. Position the plate to give the modular jack the proper vertical orientation as shown in *Figure 9-13*.

Final Check

After completing installation of all outlets and connections, it is suggested that again you run a continuity check at the cable legs A & B of the loop at the service entrance. This is the same procedure as done after rough-in of the cable runs. If continuity is obtained for each similar colored wire loop, check from one colored wire to each dissimilar colored wire. If any continuity is found, there is a short from one wire to another. Check each outlet to determine if a bare wire from one colored conductor is touching another colored conductor's bare wire. Clear up any shorts.

Call the telephone company and order your service.

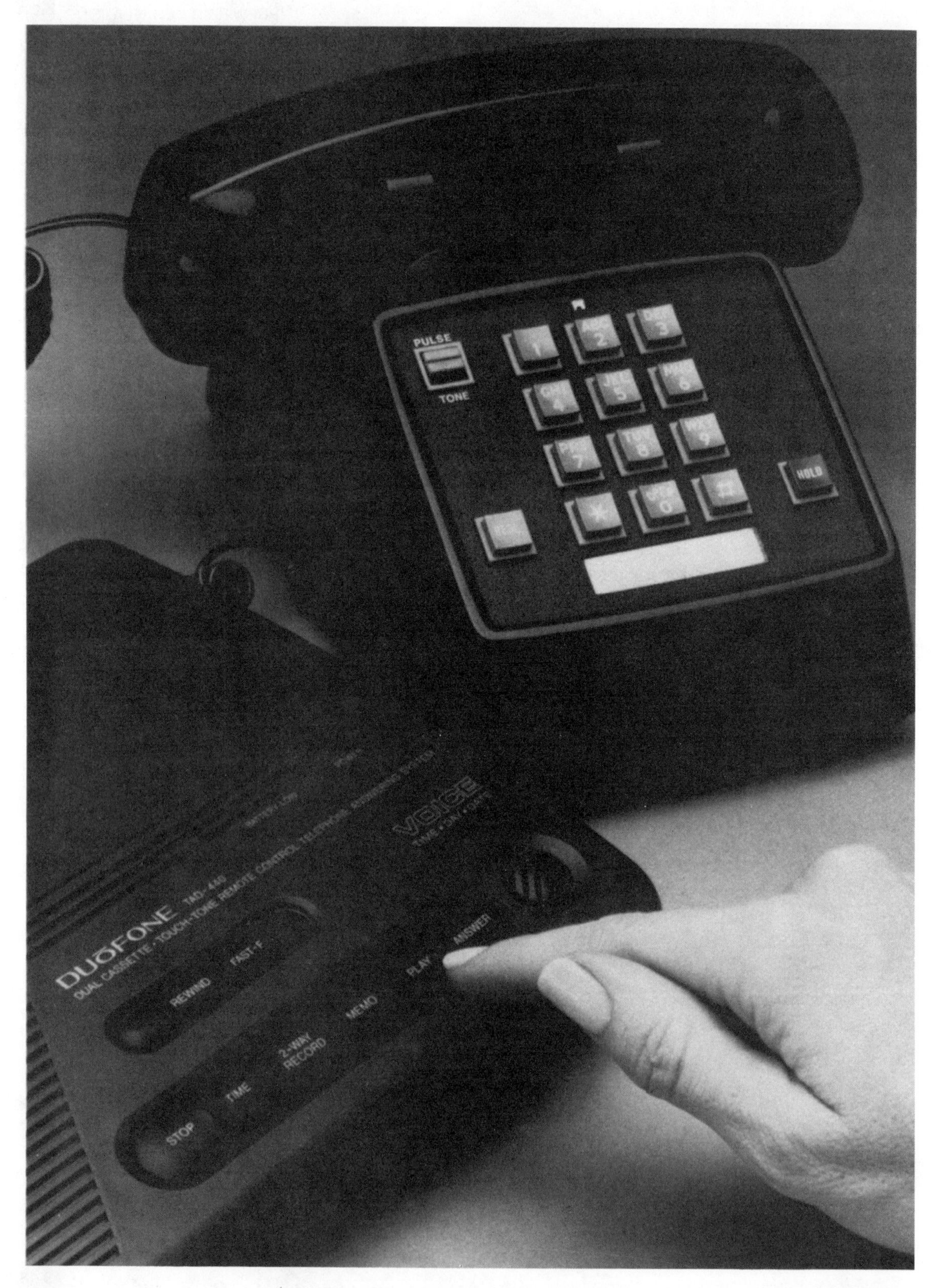
PULSE
TONE
HOLD
DUOFONE TAD-440
DUAL CASSETTE • TOUCH-TONE REMOTE CONTROL TELEPHONE
VOICE
TIME • DAY • DATE
REWIND
FAST-F
PLAY
ANSWER
STOP
TIME
2-WAY RECORD
MEMO

10-1

10 ACCESSORY EQUIPMENT INSTALLATIONS

Chapter Contents *Page*

Reminder	2
Automatic Dialers	3
How to Connect	4
Basic Types	4
Cordless Telephones	5
How to Connect	6
Basic Types	7
Telephone Answering Machines	7
How to Connect	8
Basic Types	9
Modems	10
Telephone Amplifiers	14
How to Connect	14
Miscellaneous Accessories	15
Flasher Module	15
Pocket Tone Dailer	17
Ringer Silencer	17
Outdoor Ringer	18

There are many accessories available to connect to a telephone line — auto dialers, cordless telephones, telephone answering machines, modems, amplifiers or speaker telephones, and numerous miscellaneous equipment. This chapter discusses the basic concepts of the accessory equipment, gives an idea of how it works, and explains how the equipment is connected to the telephone line.

After reading this chapter you should:

A. Know the basic concepts of the above mentioned accessory equipment that may be connected to a telephone line.

B. Have an idea how the accessory equipment is connected to the phone line.

C. Know something about choosing accessory equipment that will be most cost effective for your applications.

REMINDER

Before Connecting Any Accessory

Call your local telephone company before connecting any accessory and give the following information as you did for the telephone you purchased:

FCC Registration No.________________
Ringer Equivalence________________

In addition, the following information usually is listed:

Manufacturer________________
Model________________
U.S.O.C. No.________________
Date of Manufacture________________

The FCC does not require some accessories to have a registration number.

No Party Line or Coin Operated Telephone Lines

You are not allowed by FCC Regulations to connect any of your own purchased telephones and accessories to party lines or lines that are connected to coin operated telephones.

AUTOMATIC DIALERS

How They Work

As shown in *Figure 10-1,* auto dialers are basically electronic memories that store telephone numbers. The numbers are stored as digital codes of bits, and when requested by the press of a button, the codes are converted to a pulse-dialed or tone-dialed telephone number and sent over the telephone line. Pulse dialing allows use of both pulse-dialing and tone-dialing lines, tone dialing only on tone lines.

Many auto dialers need batteries that must be installed separately to maintain power on the electronic memories because the memories will lose the telephone numbers stored in them if the power fails. Other models have built-in protection so that the memories maintain the stored numbers.

Multiple buttons are available on auto dialers. In *Figure 10-1,* each button represents the address of the location in memory where a telephone number is stored. Most automatic dialers can store from 20 to 100 telephone numbers and can handle from 28 to 32 digits. Two keys must be used to specify the memory location; however, priority keys give one-button dialing for emergency or frequently dialed numbers. Some units provide the capability to place a two to four second pause in memory, and to combine numbers at different memory locations for chain dialing. An LTD key also is available on many units for switching to tone dialing to access special data and communications services. With those features, one is able to dial international telephone numbers that have a large number of digits. Also with the pause feature, it is possible to pause for the dial tone when using auto dialers to dial 9 to access an outside line from a business telephone.

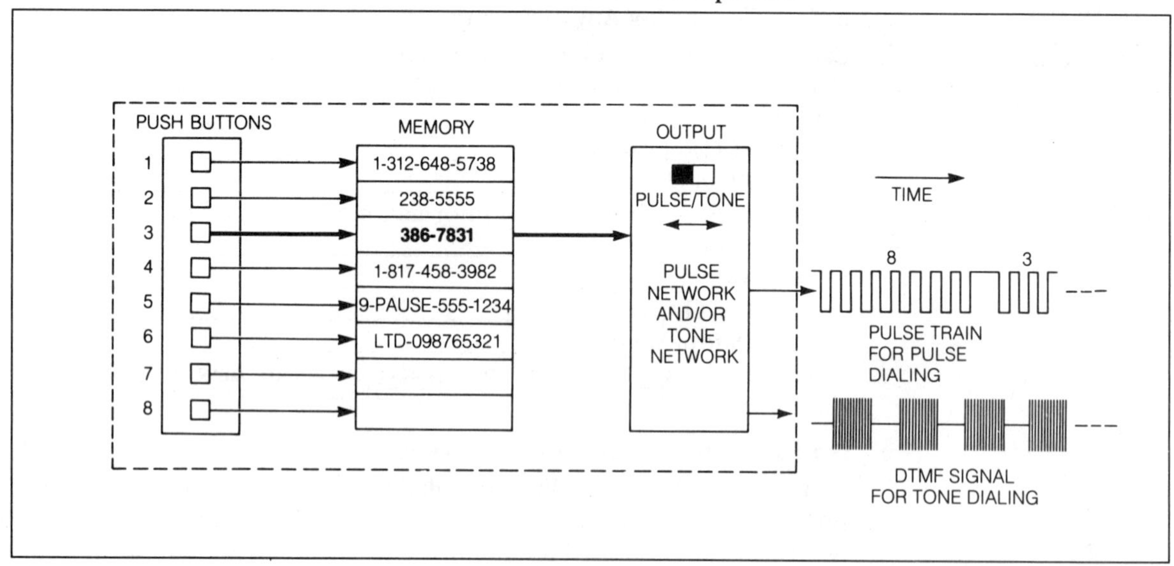

Figure 10-1.
Concept of Automatic Dialer

How Do You Connect Auto Dialers?

Connecting an auto dialer is very simple as shown in *Figure 10-2.* Unplug the existing telephone from the outlet and plug it in the modular jack in the rear of the auto dialer. Plug the modular plug from the auto dialer in the modular jack of the telephone outlet. This allows the telephone to be used normally as before, and the auto dialer now is connected to provide memory dialing.

In some cases, the auto dialers receive their power right from the telephone line; in other cases, ac power is required — either directly or through an adapter as shown in *Figure 10-2.*

Basic Types of Memory Dialers in Telephones

Advances in integrated circuit technology has allowed incorporating the memory dialing into the telephone instrument. Here are dialing features available in many combinations and on many varieties of telephones with memory dialing:

1. Names stored as well as numbers.
2. Memory storage of from 20 to 300 numbers with from 14 to 32 digits for each memory location.
3. Priority keys for frequently dialed or emergency calls.
4. Redial of last number dialed.
5. Automatic redialing of a number up to 10 times.
6. Pause key for inserting 2 - 4 second pause.
7. LDT (Long Distance Tone) key to change from pulse dialing to tone dialing when accessing special data or communications services.

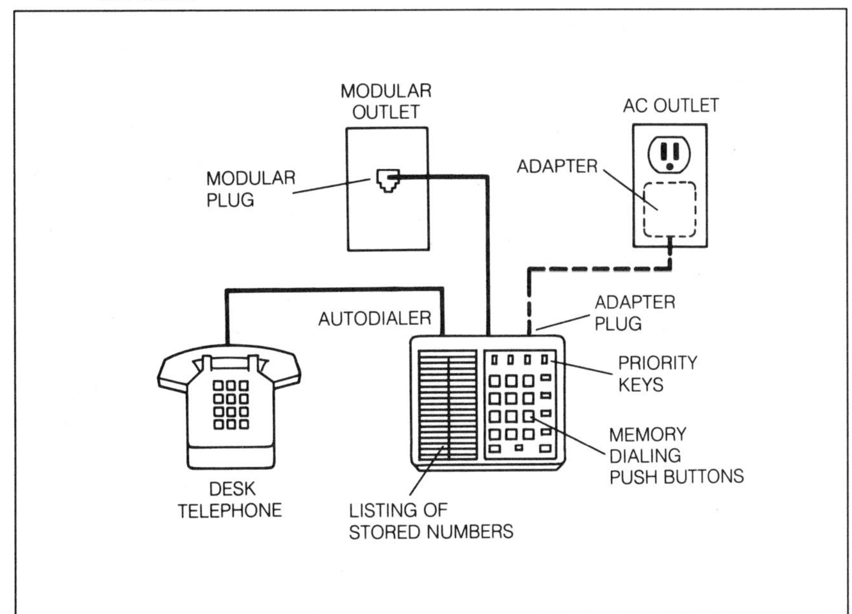

Figure 10-2.
Auto Dialer Connection

CORDLESS TELEPHONES

How They Work

Cordless telephones consist of two units as shown in *Figure 10-3*— the base unit and the handset. The base unit has two connections: a modular connection to your telephone line and a power input; either a line cord for 120 VAC or a receptacle for a voltage input from an adapter. The adapter is plugged into an ac outlet. The ac or dc voltage provides power to run the electronics and to recharge the nickel cadmium batteries for the handset electronics. The batteries are recharged when the handset is returned to the base unit. As shown in *Figure 10-4*, the base unit transmits within a frequency band of 46.6 to 46.99 MHz and receives within a frequency band of 49.6 to 49.99 MHz. The handset is just the opposite. It transmits within the frequency band of 49.6 to 49.99 MHz and receives within the frequency band of 46.6 to 46.99MHz. Telescopic antennas are built into each unit to provide the transmission and reception signals.

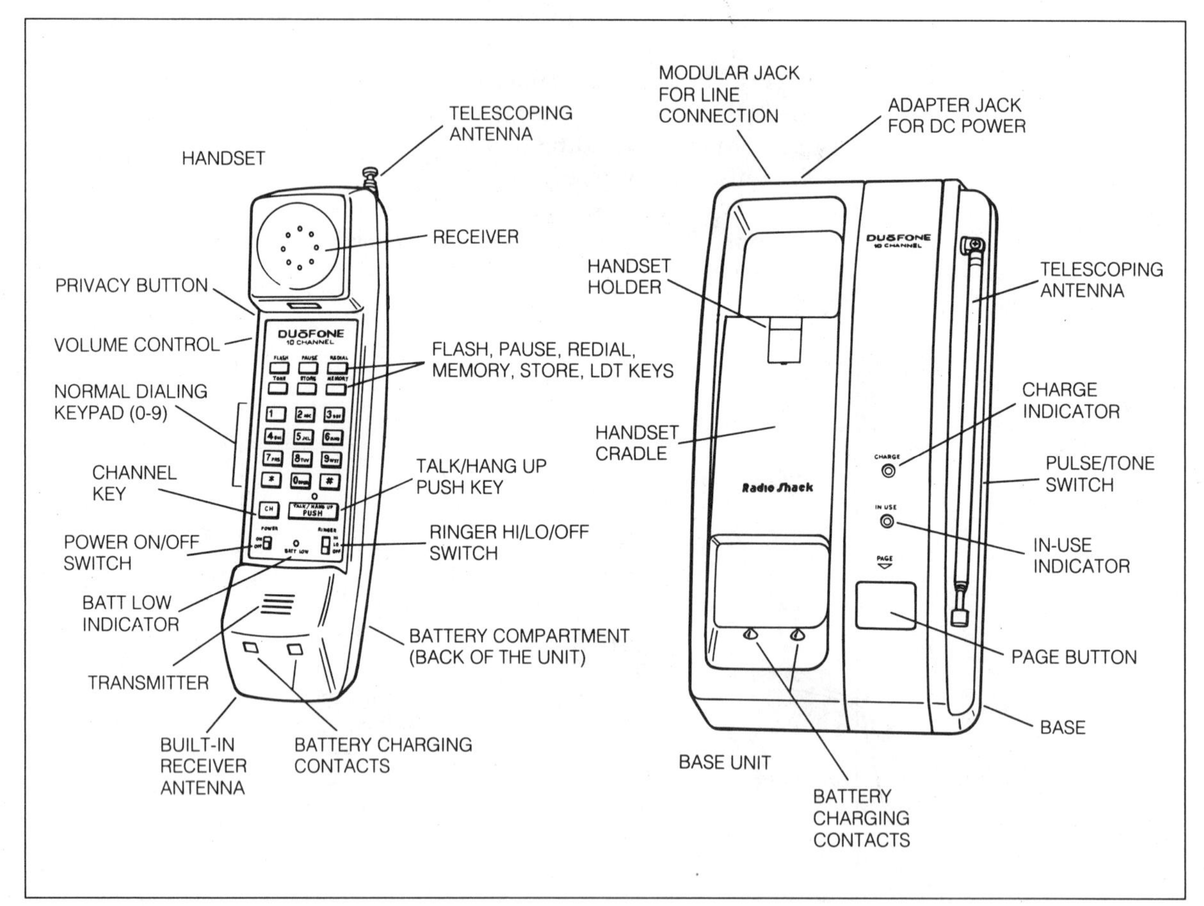

Figure 10-3.
Full-Feature Cordless Telephone
Source: Radio Shack

How Do You Connect a Cordless Telephone?

Installation of a cordless telephone is very simple. As shown in *Figure 10-4*, the base unit of the cordless telephone has two cords. One cord has a modular plug on the end. The telephone line connection is completed by plugging the modular plug in a modular outlet. The other cord is for power. If it is an ac line core, plug it into a 120 VAC outlet directly; if there is an adapter included in with the telephone, plug the cord from the adapter into the telephone base unit and then plug the adapter into the ac outlet. It is best to use a *grounded* ac outlet.

For most cordless telephones installed for the first time you will likely have to do the following:

1. *Location:* The range of the telephone is affected by the materials contained in and on interior walls — especially metal, by motors and appliances, by wireless intercoms, room monitors and alarms. Choose a location near the appropriate outlets but with open space. Reception is usually better the higher the base unit.
2. *Battery Charging:* The rechargeable battery in the handset will probably be discharged. Follow the instructions enclosed with the telephone for charging. It will take about 24 hours.
3. *Security Code:* Read the instructions carefully. Many cordless telephones require the user to set switches or insert a code for the security code. Such codes prevent other people from using your telephone and charge calls to your line.

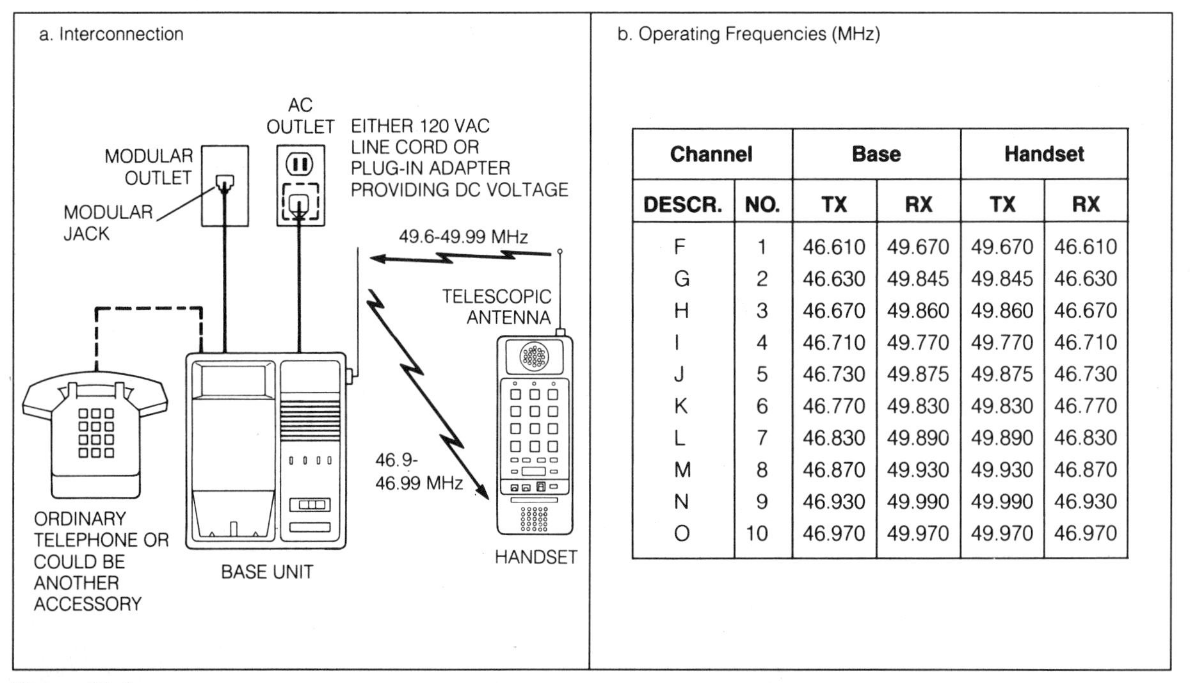

Channel		Base		Handset	
DESCR.	NO.	TX	RX	TX	RX
F	1	46.610	49.670	49.670	46.610
G	2	46.630	49.845	49.845	46.630
H	3	46.670	49.860	49.860	46.670
I	4	46.710	49.770	49.770	46.710
J	5	46.730	49.875	49.875	46.730
K	6	46.770	49.830	49.830	46.770
L	7	46.830	49.890	49.890	46.830
M	8	46.870	49.930	49.930	46.870
N	9	46.930	49.990	49.990	46.930
O	10	46.970	49.970	49.970	46.970

Figure 10-4.
Interconnection of Full-Feature Cordless Telephone Answering System

4. *Dialing Mode:* Set the pulse/tone switch to provide the dialing mode provided by your telephone service.
5. *Memory Dialing:* If you have memory dialing, you will have to store the numbers you want in memory. Check the specifications for the number of digits handled.
6. *Activate Line 2:* If the cordless telephone is a 2-line unit, Line 2 is turned off at the factory. Consult the operating manual for the switch setting to turn on Line 2 if you want to use it.

Basic Types of Cordless Telephones

Cordless telephones are available in all varieties. The most limited ones have pulse/tone dialing selection, one-button redial and a factory set security code. They have a shorter range (maybe as low as 50 ft.) and may require replaceable batteries for the handset. On the other end of the spectrum, a full-feature cordless telephone can have programmable pulse/tone dialing, at least a 30-number memory, intercom between base and handset, selectable security codes and communications channels, batteries in the handset that are rechargeable from the base, ringer control, and single-button control of pause, flash, privacy and LDT. They may have a range from 700 to 1000 feet. Either of these units, in fact, all cordless telephones will operate on at least one of the frequency channels shown in *Figure 10-4b.*

The features of six basic types of cordless telephones are shown in *Table 10-1.* Two very useful features are contained in the one-line models. Everytime the handset is returned to the base unit the security code changes automatically to a new one from 64K choices. This makes it literally impossible for someone else to use the telephone without the owner's permission. In addition, if channel interference occurs, the communications channel can be changed to any of the ten channels in *Figure 10-4b* with the press of a button. A channel that's best for the location of the telephone can be selected. Cordless telephones that have speakerphones in the base unit and ones that operate on two lines also are available.

TELEPHONE ANSWERING SYSTEMS

How They Work

Basically, as shown in *Figure 10-5,* the telephone answering system is two cassette record and playback units packaged together. One unit has a prerecorded message which it places on the telephone line when it is triggered to do so, and a second unit records a message from the calling party when it is triggered to do so.

The unit that places the message on the telephone line when the telephone is answered is called the announcement cassette and the message placed on the line is called the announcement. It usually has a time limit of from 10 to 60 seconds. When the announcement is finished, the message recorder is turned on and the announcement recorder is turned off. The calling party's message is usually limited to 30 seconds to 3 minutes. Many systems have remote control features that allow playing back recorded messages to any telephone wherever you are located. Tone-dialing telephones or a pocket tone dialer is required for the remote control.

How Do You Connect an Answering System?

Connecting a telephone answering system is very simple as shown in *Figure 10-5.* The answering system has a cord with a modular plug on the end. Plug the modular plug in a modular jack of a telephone outlet. If an existing telephone is plugged in the outlet, use a duplex jack as shown in *Figure 10-5.* The duplex jack allows the answering system and the standard telephone to be connected to the same outlet. The standard telephone can be used as before and the answering system just waits on the line until it is needed.

The electronics inside the answering system are powered either by a 120 VAC line cord or an adapter that converts the 120 VAC to a dc or ac voltage.

Cordless Telephone Features

Basic Models								Memory Dialing		Power		Security					
	Pulse/Tone	Redial	Ringer Control	Privacy	Flash	LDT	Pause	Numbers	Digits	120 V	Adapter	Code	Choices	Channels	Intercom	Recharge	Speaker Phone
A	P	●	●	●	●	●	●	30	16		●	●	*64K	10	X	●	●
B	P	●	●	●	●	●	●	30	16		●	●	*64K	10	H	●	
C		●		●	●	●	●	30	16		●	●	*64K	10		●	
D		●		●	●	●	●	3	16		●	●	*64K	10		●	
E	●	●	●		●						●	●	FS			C	
F 2-LINE	●	●		●		●				●		●	**128	3		●	

Notes:
P - One-button selectable
H - Base and handset can beep each other
C - Cordless handset uses C batteries
FS - Factory set
* - Automatically changes when handset is returned to base
** - Manual selection

Table 10-1.
Cordless Telephone Features

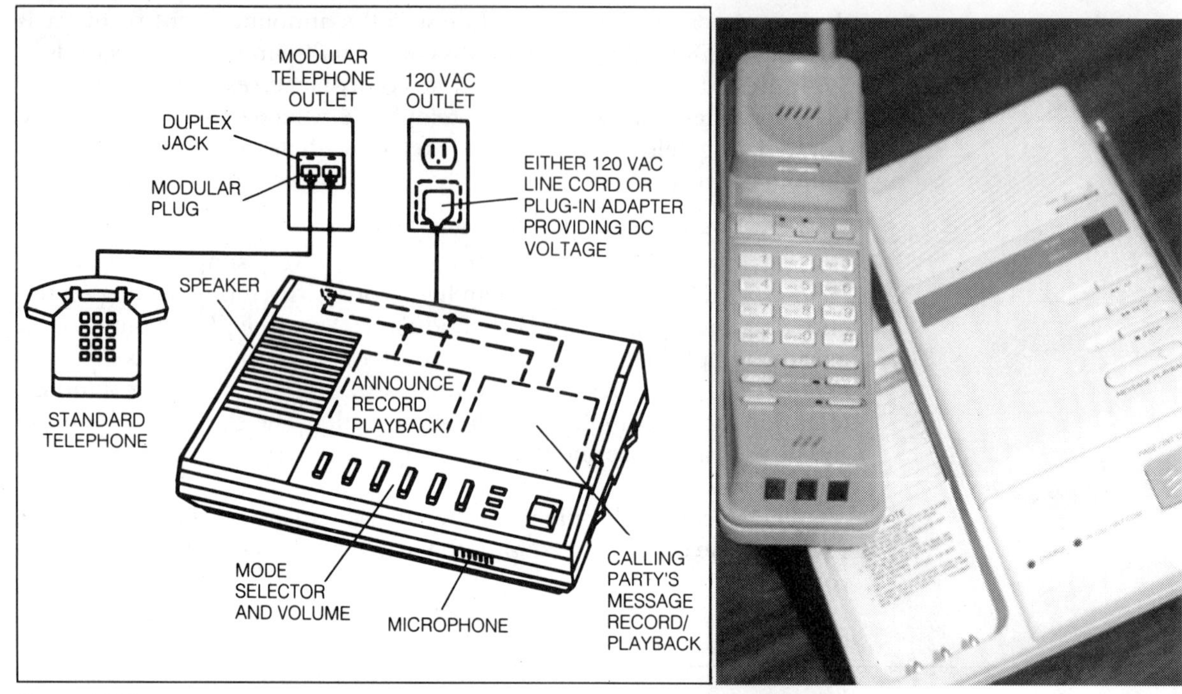

Figure 10-5.
Telephone Answering System

Figure 10-6.
Telephone and Answerer

Basic Types of Telephone Answerers

Advances in technology have also contributed to the addition of many features to answering machines. For example, as shown in *Figure 10-6*, a speakerphone and answerer are combined in one unit that has 20-number memory dialing and remote control through a tone-dialing telephone. Common features of most answerers are:

1. Sensors to detect tape jam or break and keep the telephone line free if the tape malfunctions.
2. Protection circuit to detect power interrupt.
3. Protection circuit for electrical spikes on the power line.
4. User selectable security code settings rather than inflexible factory settings.

Several common types of answering machines are on the market, including:

1. A sophisticated system that allows any length announcement message, has first-ring or delayed ringing answering, selectable incoming message length of from 30 seconds to three minutes, and an announce-only mode when no incoming messages are to be

received. The answerer is controlled by voice activation and if the calling party hangs up, the answerer stops. These features save tape and playback time. The unit can be interrogated and controlled with tone-dialing signals over a remote telephone for turn-on, cue/review, rewind and erase, and has a security code and a toll saver feature — if any new messages have been recorded, it answers on the first ring, otherwise it will not answer for several rings so the caller has the opportunity to hang up.

2. A full-feature telephone answerer shown in *Figure 10-7* has all the features of the sophisticated model of item 2 plus additional features. It provides call monitoring of incoming messages, the ability to remotely monitor the room where the answerer is located, gives helpful verbal instructions for operating the machine, announces the number of calls received and messages left, stamps the time and date on each incoming message, and records 2-way conversations.

And 2-line models are available with many of these single-line features.

MODEMS

When you are talking on the telephone, the electrical signals that are transmitted are called "voice signals" because they only contain signal frequencies that are typical for voice conversations — from 200 to 6,000 Hz (cycles per second). The common telephone line over which the voice signals are transmitted does not pass the full band of voice frequencies. It restricts the signal to a band of frequencies between 200 Hz and 3,400 Hz.

Figure 10-7.
Full-Feature Telephone Answerer
Courtesy of Southwestern Bell Freedom Phone®

Data Transmission

If one computer or data terminal wants to send information (data) to another computer or data terminal, the information is in the form of a digital code that contains frequencies that are much higher than the band from 200 Hz to 3,400 Hz. Therefore, the computer data cannot be transmitted over standard telephone lines unless it is converted to a signal that is within the voice frequency band.

Purpose of a Modem

The purpose of a modem is to perform the signal conversion. It converts the digital data signals into electrical tone signals that are within the voice band frequencies. The conversion of the digital data signals into tone signals with frequencies that are within the voice band is called *modulation.* The modulation occurs at the computer or data terminal that is sending or *originating* the information.

When the data arrives at the receiving computer or data terminal, in order for it to be recognized by the computer or data terminal, the data must be converted back to a digital code. Converting the tone signals in the voice frequency band back to a digital code is called *demodulation.* Demodulation occurs at the receiving or *answering* computer or data terminal.

A modem performs either the modulation or demodulation depending on the end of the telephone line to which it is connected. In fact, that is how the modem got its name. MODEM stands for MOdulation and DEModulation.

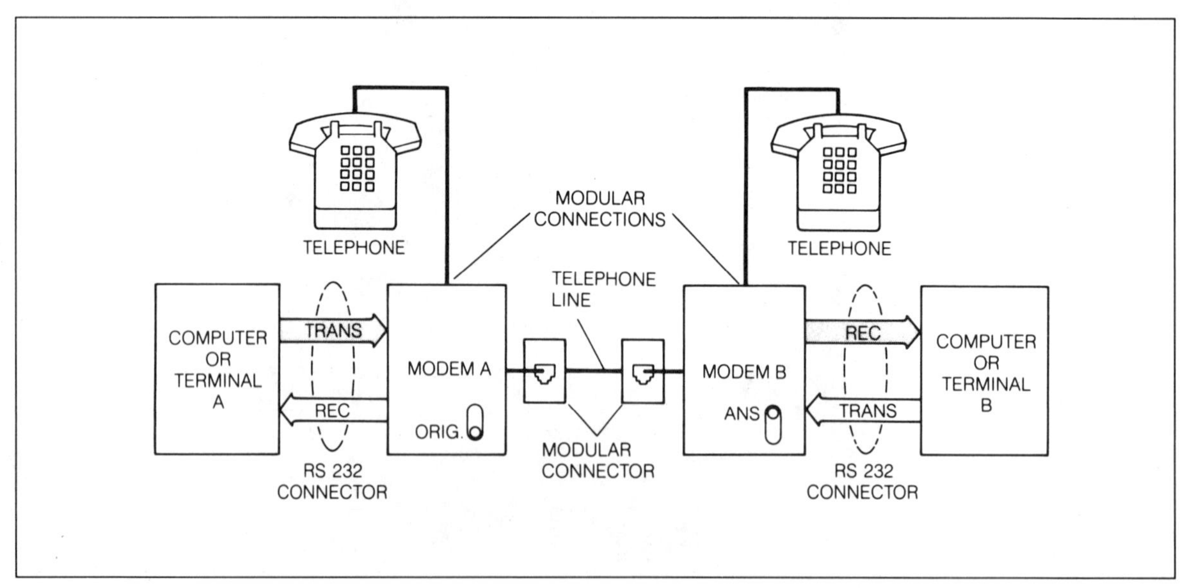

Figure 10-8.
Modem Interconnection

How a Modem Works

Figure 10-8 shows the interconnection of computers A and B (or data terminals) through modems to the telephone line. If computer A is sending data to computer B, computer A is transmitting; modem A is in the *originate* mode and modulating; modem B is in the *answer* mode and demodulating and computer B is receiving. In order to make sure the connections are correct and each system is recognizing the signals properly, computer B sends back a signal to computer A — it *answers* computer A. For this reason, it is in the *answer* mode. When computer B sends back the answer signal, it is transmitting; modem B is modulating; modem A is demodulating and computer A is receiving. Signals are going from computer A to computer B when computer A is originating the signal and from computer B to computer A when computer B is sending an answering signal.

Modulation

In *Figure 10-8,* modem A is in the originate mode. The digital code from computer A is called a binary code because it has *two* levels as shown in *Figure 10-9a.* The two levels generate two tones out of the modem as shown in *Figure 10-9a* and *10-9b.* One tone is called a *mark;* the other tone is called a *space.* When in the originate mode, the mark tone sent out

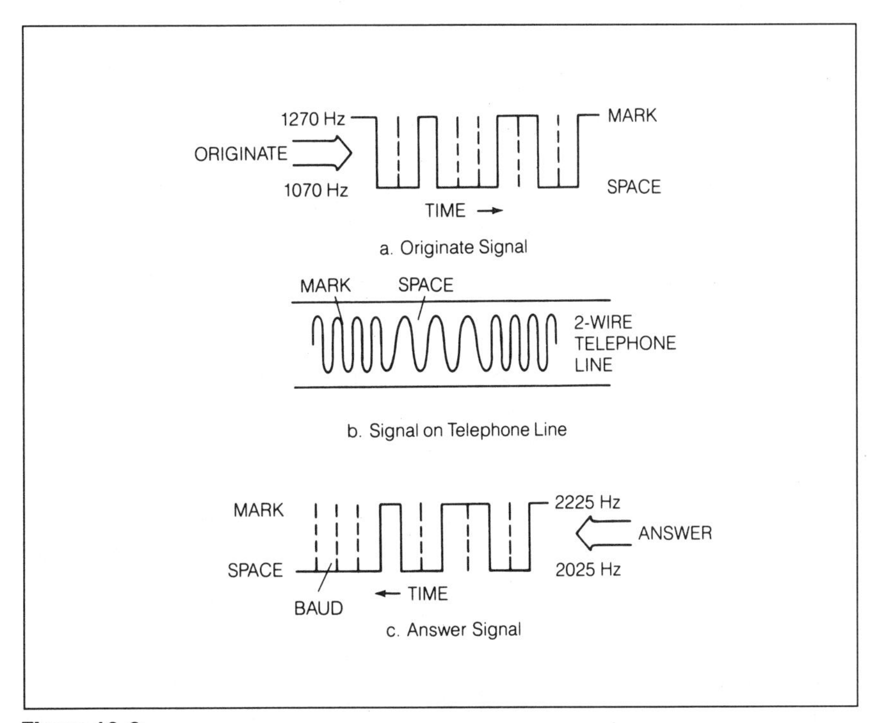

Figure 10-9.
Modem Signals

by modem A is 1270 Hz. The space tone is 1070 Hz. Therefore, instead of a digital signal that has many high frequency tones, every time the digital code from computer A varies between its two levels, either a 1270 Hz tone (mark) or a 1070 Hz tone (space) is sent down the telephone line. Each of these signals is within the voice frequency band. The telephone line can transmit these tones very well because they are just like voice signals.

When computer B answers computer A, the two tones for the mark and space of the digital code that modem B uses for its modulation are different from those used by modem A. As shown in *Figure 10-9c*, the mark tone is at 2225 Hz and the space tone is at 2025 Hz. The reason for this is so that the computers can recognize each other's signals because the signals can be on the line at the same time. In technical terms this is called *duplex operation* because transmitting and receiving can go on in both directions at the same time.

Demodulation

When the mark and space tones sent by computer A arrive at computer B, modem B demodulates the tones and restores the binary digital code and sends it to computer B. In like fashion, modem A demodulates the 2225 Hz mark tones and 2025 Hz space tones sent from computer B to computer A and converts them to a binary digital code that is received by computer A.

Baud Rate

As shown in *Figure 10-9c*, the time that is used to generate a mark or a space is called a baud in technical terms. It is not very important to you except to know that modems are rated in how many bauds per second they can transmit and receive. Common baud rates for modems are 300 to 9600 bauds per second. Obviously, it takes much less time to transmit data at 9600 bauds per second than it does to transmit it at 300 bauds per second.

Half-Duplex or Simplex Operation

When transmitting and receiving only can occur in one direction at a time, it is called simplex or half-duplex operation.

RS-232

The digital code signals from the computers or data terminals to the modem must meet certain electrical values and be arranged in a particular way and occur in a particular time relationship. A written explanation of the relationships is called a specification or in more technical terms a *protocol.* One such specification or protocol is called RS-232.

When manufacturers of equipment tell you that a signal from digital equipment meets RS-232 specifications then you know that you can connect it to another piece of equipment that uses RS-232 and it will work properly.

How Do You Connect a Modem?

Figure 10-8 shows a typical data communication system. A modular plug from each modem plugs into a telephone line outlet and the telephone that was plugged into the outlet is plugged into the modem. The modem completes the connection from the telephone set to the telephone line and may disconnect the telephone set from the line when the modem is modulating or demodulation.

The computer or data terminal connection is made by plugging together standard RS-232 25-pin connectors used by computer and data terminal manufacturers. In many cases, the modem is on a card that plugs into the computer and the connections are internal. Determine if your modem is going to transmit or receive. If it is going to transmit, your modem should be in the originate mode and the receiving modem in the answer mode. If it is going to receive, then your modem should be in the answer mode and the transmitting modem in the originate mode. In *Figure 10-8,* modem A is originating and modem B is answering.

The function of the telephone set in *Figure 10-8* is to provide the connection through the telephone switching network from the calling party (modem A) to the called party (modem B). In other cases, the telephone handsets must be placed in foam rubber cups on the modem to acoustically couple the modem signals to the telephone line.

Most modems have a standard plug for an ac outlet or else are powered from an adapter supplied with the unit.

TELEPHONE AMPLIFIERS

How Does a Telephone Amplifier Work?

Telephone amplifiers are of two types — directly coupled electrically or indirectly coupled acoustically or magnetically. For acoustically or magnetically coupled, the sound waves strike a diaphragm of a microphone which converts the sound waves to electrical signals just like the transmitter of the telephone handset. An electrically coupled amplifier will get its signals directly from the telephone line without any microphone. It will amplify the electrical signals and deliver a higher level signal to a speaker. At the speaker, the electrical signals are converted to sound waves.

How Are Different Amplifiers Connected?

Figure 10-10 shows the interconnection of various types of amplifiers within the telephone system. The A amplifier straps to the receiver and input signals are magnetically coupled from the receiver. Signals coming from the receiver are amplified 5 or 10 times by the strap-on amplifier and are output as higher-level sound waves. There are no electrical connections to be made, and no registration is required with the telephone company.

Amplifier B shown in *Figure 10-10* also does not require registration. It only amplifies the incoming calls and does not have a built-in microphone and the necessary switching for a two-way amplifier. It receives its input signals from a pick-up coil that is attached to the telephone with a suction cup.

With full-feature two-way amplifiers or speaker telephones like that shown for amplifier C in *Figure 10-10*, the telephone connects into the amplifier and the amplifier connects to the telephone line. As a result, you must register the amplifier with your local telephone company.

As shown in *Figure 10-10*, to connect amplifier C, disconnect your telephone modular plug from the telephone outlet and plug it in the modular jack on the amplifier. Plug the modular plug from the amplifier in the telephone modular outlet.

The amplifier usually receives its power from the telephone line. If not, an adapter or an ac plug will be provided. It will have an ON/OFF switch, a volume control and may have a privacy (mute) switch. Usually the microphone is on the front of the unit and the speaker is behind a grill or cloth mesh.

MISCELLANEOUS ACCESSORIES

A telephone flasher module, a pocket tone dialer, a ringer silencer and an outdoor ringer are additional accessories that may prove useful.

Telephone Flasher Module

A new unique function, especially useful for people with impaired hearing, is available for home or business that can be installed on single party lines. It is called a telephone flasher module. It is connected to the

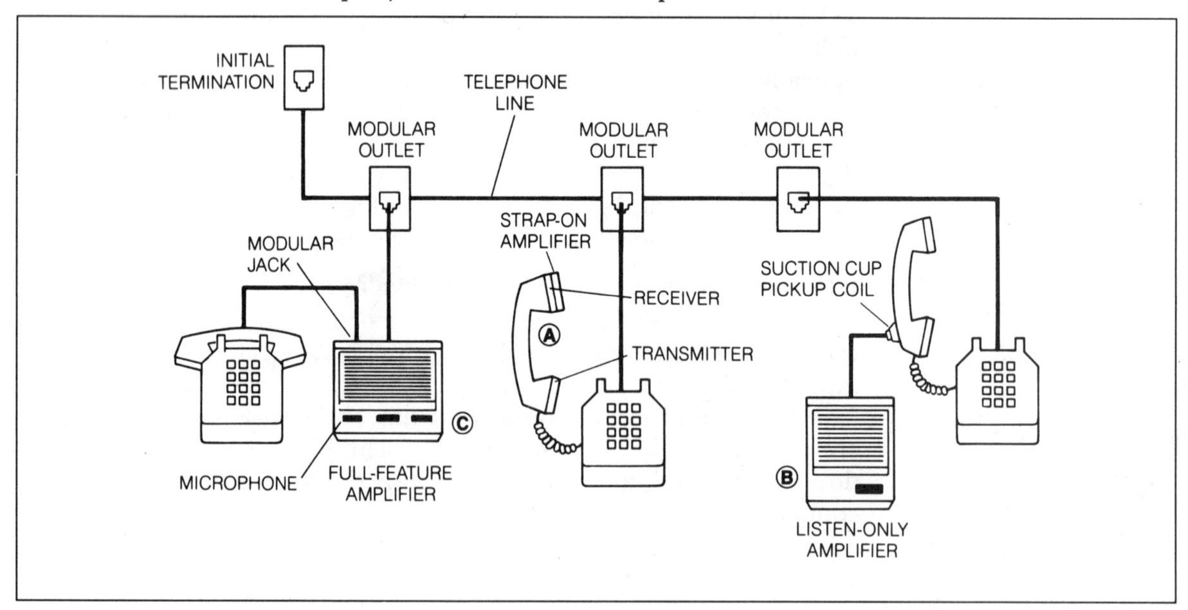

Figure 10-10.
Telephone Amplifier Interconnections

incoming telephone line in a parallel with your telephones. When telephone ringing voltage appears on the line to ring your telephone, the flasher module discussed in Chapter 9 either flashed a strobe light contained on the unit, or sounded a loud buzzer, or did both. The flasher shown in *Figure 10-11* detects the ring signal and turns on and off a table lamp, utility light, or special signalling light that is plugged into the flasher. Circuitry inside of the flasher module recognizes the present of ringing voltage and controls power to the ac outlet for the light that is flashed.

Connecting a Flasher Module

As shown in *Figure 10-11*, the flasher module is plugged into an ac outlet for its power and its modular plug must be connected to a modular outlet to be able to provide the flasher function. Use a duplex jack to connect a telephone and the flasher module into the same modular outlet.

An incandescent lamp is plugged into the flasher module ac outlet on the bottom of the module. The flasher module is designed only for incandescent lamps. Power of up to 300 watts can be supplied, but never connect a fluorescent light or any type of appliance to the flasher outlet.

Be sure the lamp that is plugged into the flasher is turned on, otherwise the light will not flash when the telephone rings. Also, an ac outlet must be available near a telephone outlet.

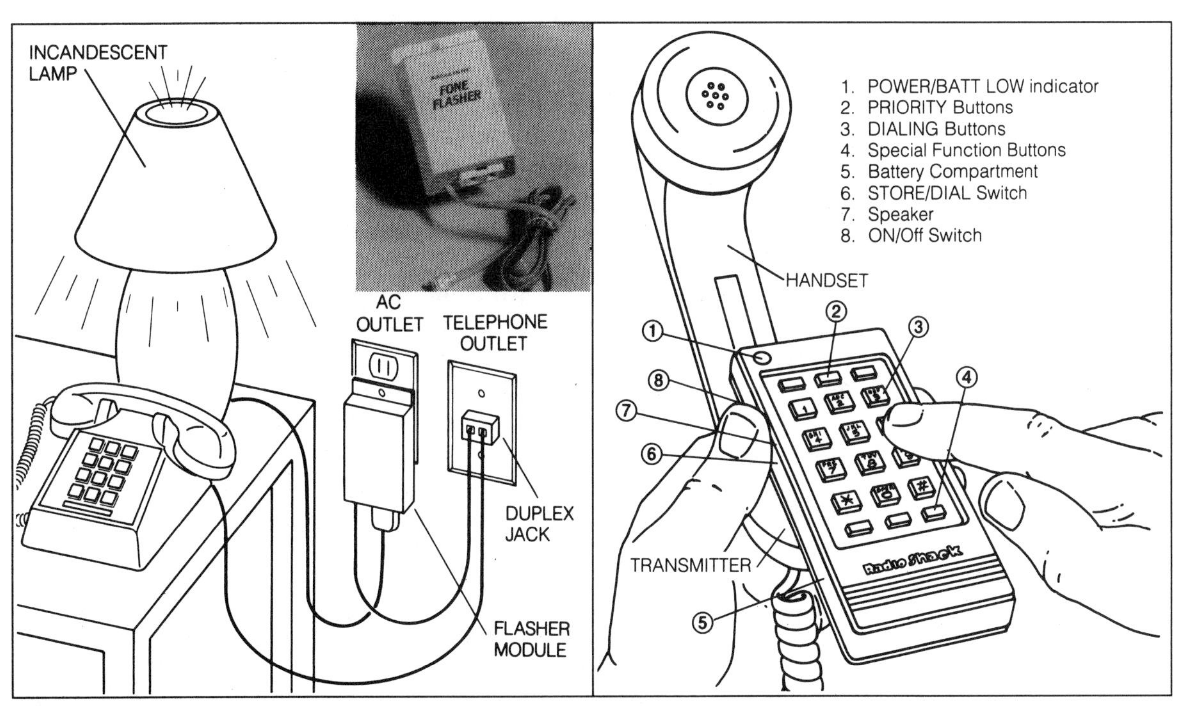

Figure 10-11.
Flasher Module
Source: Radio Shack

Figure 10-12.
Pocket Tone Dailer
Source: Radio Shack

Pocket Tone Dialer

Many times a person needs tone dialing to dial into alternate long distance service; or a salesman needs to access computer based services such as order entry, customer service inquiries, central dictation, or call diverters, and the only telephones available are pulse-dialing telephones. One easy solution is to use a pocket tone dialer. With the pocket tone dialer, tone dialing capability is always available.

As shown in *Figure 10-12,* the pocket tone dialer is a small portable unit that has a keypad the same as a telephone, It operates from its own batteries. Its output is a standard tone dialing dual-tone multifrequency (DTMF) audio sound from a self-contained loudspeaker. The audio output is coupled by sound waves to the transmitter of the telephone handset by holding the pocket tone dialer speaker tightly to the transmitter. Pressing the appropriate keys generates the correct dual tones that accomplish the dialing. It may have memory or priority number dialing.

A telephone in the pulse mode is used on the pulse-dialing line to make the normal connections from calling party to called party. After the connection is made, the tones generated by the pocket tone-dialer provide the tone-dialing information to complete the required addressing. Follow the manufacturer's instructions carefully when holding the tone-dialer to the telephone handset.

Because the pocket tone dialer couples to the telephone with sound waves (acoustically), no FCC registration number or REN number need be supplied to the local telephone company.

Ring Silencer

The easiest way to install a ringer silencer on a full modular telephone is to replace the line cord that connects the telephone set base to the modular outlet with a ring silencer cord. It has a switch in a line cord with modular connectors on both ends. Just plug the modular plugs of the cord in the telephone set base and in the modular outlet and the silencer is connected.

If your telephone is not full modular but has a line cord with no modular connector at the telephone set base, then use a line cord inline coupler as shown in *Figure 10-13.* Remove the existing line cord modular plug from the outlet and connect it to one side of the inline coupler. In the other side of the coupler plug one end of the ringer silencer cord. The other end of the ringer silencer cord plugs into the outlet.

Some units, as shown dotted in *Figure 10-13,* mount right on the telephone set and connect to the inside of the telephone set using easy instructions.

CAUTION— If a ring silencer is used, the telephone that it is used on is disabled. It will not ring or it can't be used for making a call. Use some means to remind you that the telephone is disabled. Take the handset off hook, turn it upside down on hook, or use some other physical sign that the telephone is disabled so that the telephone may be put back in service as soon as the silence period is over. Of course, in an emergency it requires only the flick of a switch to get back in service.

You do not have to notify your local telephone company when you install a ringer silencer.

Outdoor Ringer

In order to hear your telephone ringing when you are in the garage, outdoors working, or cooking on the patio, an outdoor ringer with adjustable volume is a very useful accessory. It is a large telephone bell (4" gongs) that is energized by the ringing voltage. As shown in *Figure 10-14*, all that is required to install an outdoor ringer is a modular outlet for your telephone line and a short modular-to-modular adapter cord.

Run the telephone lines using the techniques for Chapters 6 and 7. If the outlet is in the garage, use a surface-mount jack. If the outlet is outdoors, use a weather-proof outlet as shown in Chapter 9 and mount the outlet in a protected location. Even though the ringers are built for outdoor installations, some protection will extend the useful life. Mount the ringer vertical so it is protected by the plastic cover. Choose a protected place under the roof overhang (soffitt) if possible, but in a location where the sound can be heard easily.

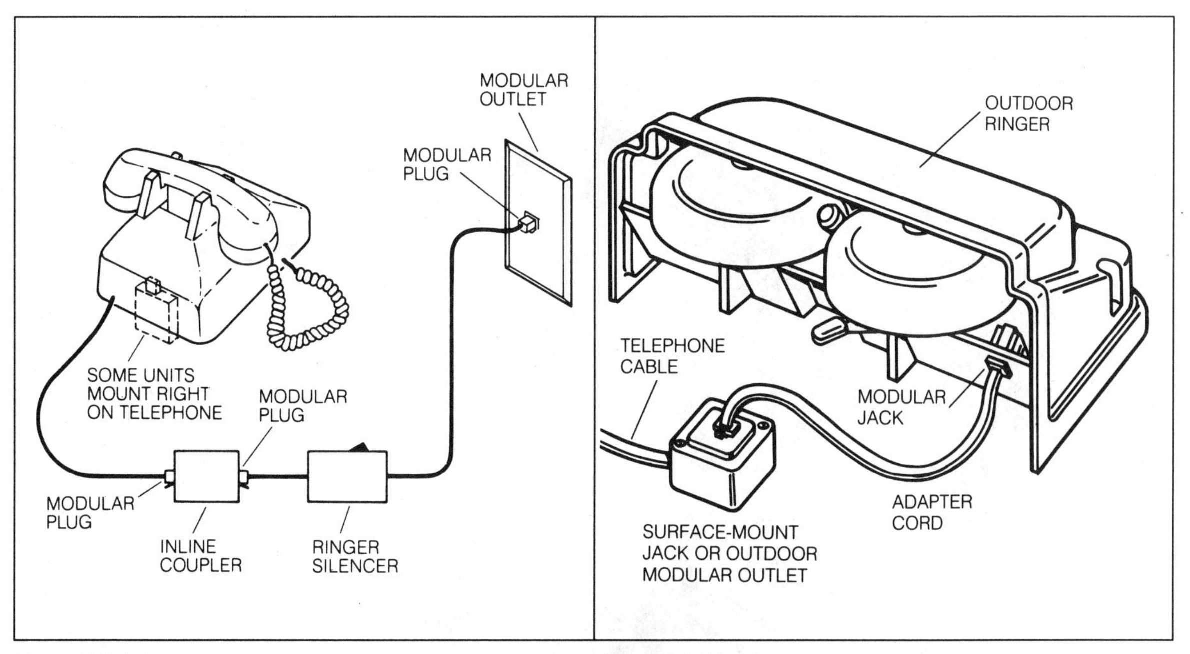

Figure 10-13. Ringer Silencer

Figure 10-14. Outdoor Ringer

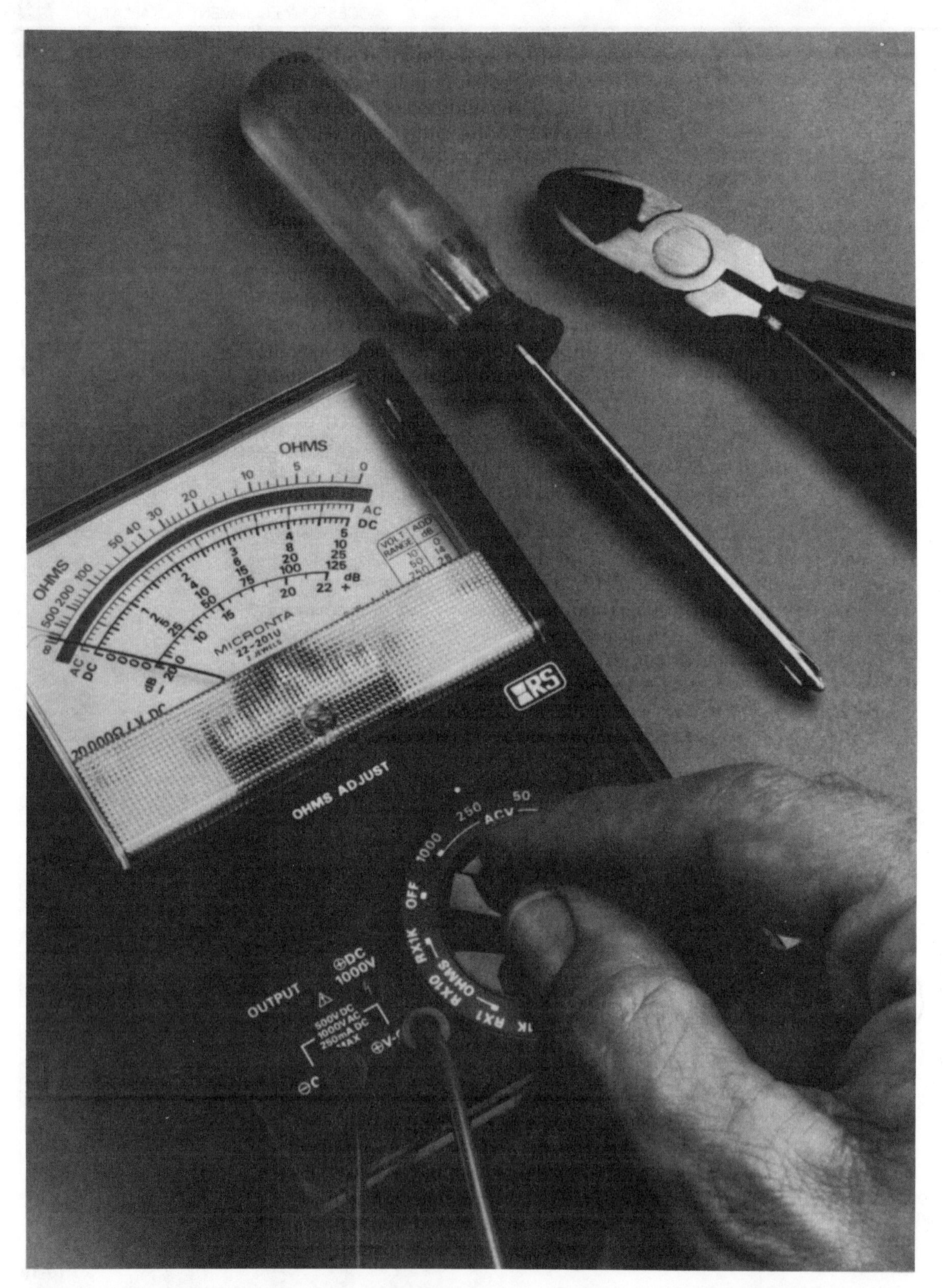
OHMS
AC
DC
MICRONTA
22-201U
RS
OHMS ADJUST
ACV
250
50
1000
OFF
RX1K
RX10
RX1
OHMS
OUTPUT
⊕DC
1000V
500V DC
1000V AC
250mA DC
MAX

11 INSTALLATION CHECKS AND TROUBLESHOOTING

Chapter Contents — *Page*

Initial Installation Checks 2
Detecting Wiring Problems 5
Isolation, Substitution and Elimination 5
Battery and Buzzer 6
Multi-Meter 8
Using Extra Cable Pairs 9
Other Common Problems 9
Modular Connectors 9
Intermittent Contacts 10
Transmitter 11
Receiver 11
Rotary Dials 11
Ringer 11
Troubleshooting Table 12

In all installations it is an objective that everything will be done correctly and all replacements and additional telephones will work properly. However, problems can arise. This chapter discusses how to deal with common installation problems and troubles that can occur in the telephones or in the wiring in your home, apartment or small business. *Remember you are not allowed to do any repair work on equipment owned by the local telephone company.*

After reading this chapter you should:

A. Be able to check your initial installation of telephones or wiring to assure that everything is working properly.

B. Be able to construct several simple test adapters to aid in locating problems.

C. Analyze problems that occur in your telephone system and repair them yourself to save the service charge.

INITIAL INSTALLATION CHECKS

Check Your Purchased Telephones

Any purchased telephones should be checked at the store where they are purchased if at all possible. The easiest way to do this is to plug the telephone in an outlet and see if it will dial some common available numbers. Time and temperature numbers are good examples.

If there are no facilities to do this, as soon as you can after purchase, substitute the telephone for one that is working in your home, apartment, or small business. When the telephone is plugged in and the handset is lifted to go off hook, a dial tone should be heard. With the dial tone present, dialing numbers will assure you the telephone is operating properly.

Tone-Dialing Telephones

Recall if your new purchased telephone is a tone-dialing telephone and you only have pulse-dialing service, it will not work properly until you call your local telephone company and request tone-dialing service.

However, another problem may arise even though tone dialing service has already been connected by the local telephone company. Your tone-dialing telephone will not dial. When the handset is off hook a dial tone is heard, but it remains on when dialing starts. The problem is solved by reversing the red and the green wire connections either at the telephone outlet or at the initial termination from the protector. Somehow, somewhere along the line the connections were reversed.

Shorts and Opens in Wiring

Two of the most common types of problems that arise in new telephone installations are an open line or a shorted line at one of the outlet connections. These two types of problems are demonstrated in *Figure 11-1* which shows a typical installation. Point MI is a modular interface jack (RJ11W or RJ11C) where the telephone line into the building is terminated initially. A telephone wiring block is connected to it with a modular plug. This is the recommended (and maybe required) connection because the new installation can be disconnected from the initial telephone company termination. This isolates the new installation in case trouble exists on the line.

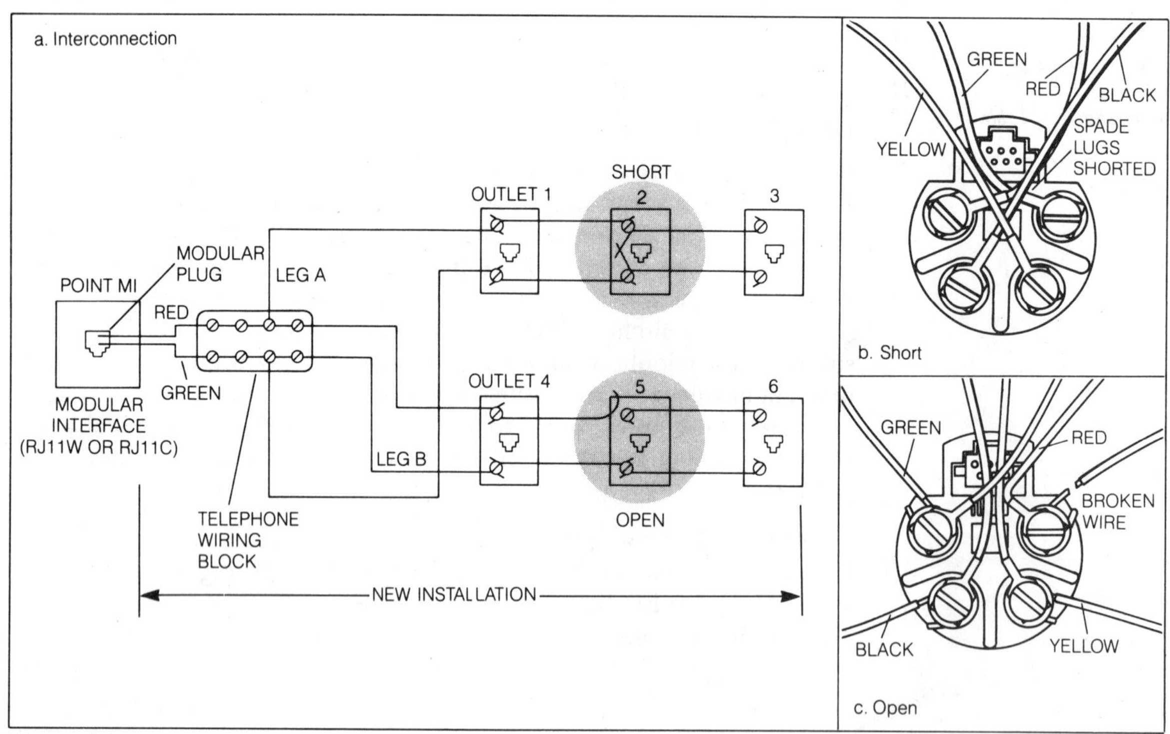

Figure 11-1.
Short and Open in Initial Wiring

Shorts

A very common occurance if one is not careful in wiring is that two bare wires wrapped around adjacent screw terminals touch each other and short the line together. This also occurs easily when spade lug terminals slip as the screw terminals are tightened and push against each other to short out the line as shown in *Figure 11-1a.*

When the telephone line is shorted at an outlet (outlet 2) as shown, the telephone that is plugged in outlet 2 will not work; neither will any of the telephones that are plugged in outlets 1, 3, 4, 5 and 6 because the voltage from the central office that produces a current in the local loop to the central office is shorted out. Current will flow continuously and it will appear to the central office that a telephone is off hook all the time. This problem will be detected at the central office, and the telephone company will send out a serviceman to correct the problem or disconnect your service.

Examine your wiring carefully as you make your installation to keep from having wires shorted together.

Opens

The other common problem is to have a bare wire break at the screw terminal to cause the telephone line to open. As shown in *Figure 11-1b,* the bare wire has broken where the insulation was stripped because the conductor was nicked when the insulation was cut. This opens the connection and any telephones connected to outlets 5 and 6 will not work because the current flowing in the local loop can not flow to the telephones at outlets 5 and 6 (this, of course, assumes that the short in the line at outlet 2 has been removed). With the short removed at outlet 2, all the remaining outlets (1, 2, 3 and 4) will provide proper telephone service when telephones are plugged in their outlets. It is important to note that *shorts prevent all outlets from working properly, while opens may prevent only selected outlets from working properly*. Of course, if there were an open at point MI of the modular interface, it also could prevent all outlets from working properly.

Another thing that can occur to produce an open line is for the wire to be stripped improperly. If the insulation is not cut back far enough when the wire is screwed down under the terminal, the screw presses on the insulation and not on the bare wire . An open line is produced because the screw terminal is not connected to the bare wire but is insulated from making the connection. This may be a difficult problem to find; therefore, connect your wiring carefully.

Examine your wiring carefully as you do your installation to make sure there are no broken wires or bad connections that produce an open line.

DETECTING WIRING PROBLEMS

Isolation, Substitution and Elimination

One good way of detecting wiring problems is to isolate the problems by the process of substitution and elimination. For example, when a telephone is plugged in any one of the outlets of *Figure 11-1*, it doesn't work. You know the telephone works because it was tested at the store where it was purchased. To isolate the problem, the modular plug at point MI is disconnected and the telephone plugged in at point MI. You have substituted the single telephone for the rest of the line. At this location the telephone works. A dial tone is produced when the handset is off hook and numbers can be dialed. As a result, the problem is in the wiring of the outlets of Leg A and Leg B beyond point MI.

Isolating Leg B

To continue the isolation, leg B is isolated by eliminating leg A. Leg A is eliminated by disconnecting leg A at the telephone wiring block. This is shown in *Figure 11-2*. The modular plug is reconnected at point MI and the telephone substituted in outlet 4. Now the telephone works — a dial tone is heard when the handset is off hook and it dials correctly.

Plugging the telephone in outlets 5 and 6 shows it still does not work in these outlets. However, knowing that the outlets 5 and 6 are fed from outlet 4 has isolated the problem to outlets 5 and 6.

The wiring at outlet 5 is examined closely and the open wire detected and repaired. The telephone is plugged in outlet 5 and it works and in outlet 6 and it works. The problem was isolated to leg B and to outlet 5 by the process of elimination and by substituting a good telephone in outlets 4, 5 and 6.

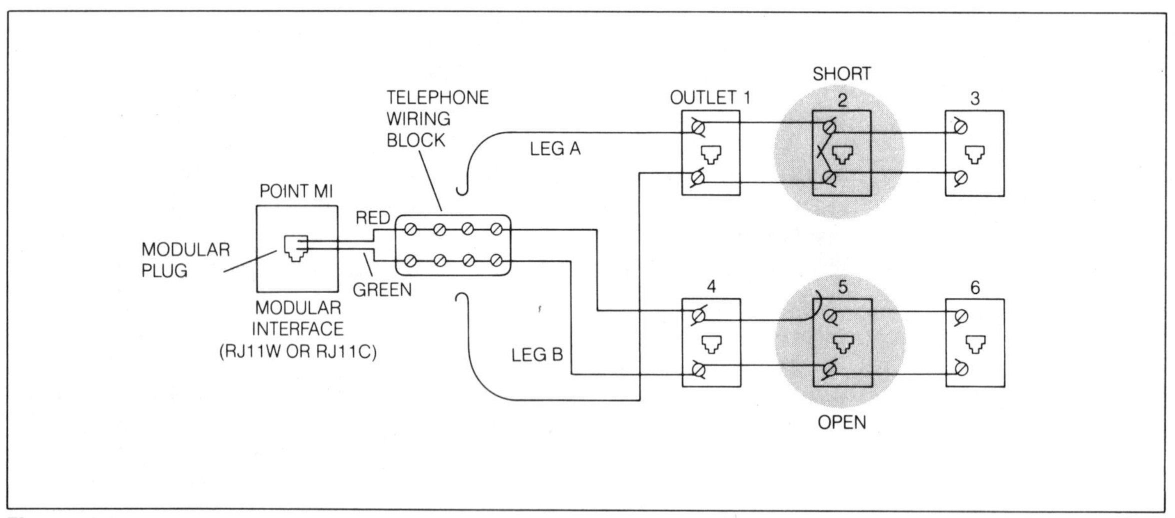

Figure 11-2.
Leg A of Figure 11-1 Eliminated

Isolating Leg A

To continue the isolation, with the open repaired in leg B, leg A in *Figure 11-2* is reconnected at the telephone wiring block. Plugging in the telephone at outlets 4 and 1 shows again that the telephone doesn't work in either outlet. A problem still exists in leg A. The outlet at 3 is examined carefully and its wiring is ok — no opens, no shorts. *Figure 11-2* shows that outlet 3 is fed by outlet 2, so the outlet 2 wiring is examined next. Careful examination shows that the two wires are shorted on outlet 2. The short is removed and the telephone now works when plugged in outlet 2 — a dial tone is heard when the handset is off hook and it dials correctly.

To complete the troubleshooting, outlets 1 and 3 now provide proper operation when the telephone is plugged in them. Thus, by a process of elimination by substituting a good telephone and isolating the problem, the short in leg A and the open in leg B have been repaired. The complete system now works properly.

Battery and Buzzer

Another way of detecting wiring problems is to use a continuity checker made from a battery and a buzzer as shown in *Figure 11-3*. Continuity means that a complete circuit is formed and current supplied by the battery flows in the circuit and energizes the buzzer. Assume that the same wiring problems exist as for *Figure 11-1*. The battery and buzzer can be constructed from a 9V transistor battery with snap connectors, a buzzer and a set of wire leads with alligator clips on the ends. Use some of the telephone cable for the wire.

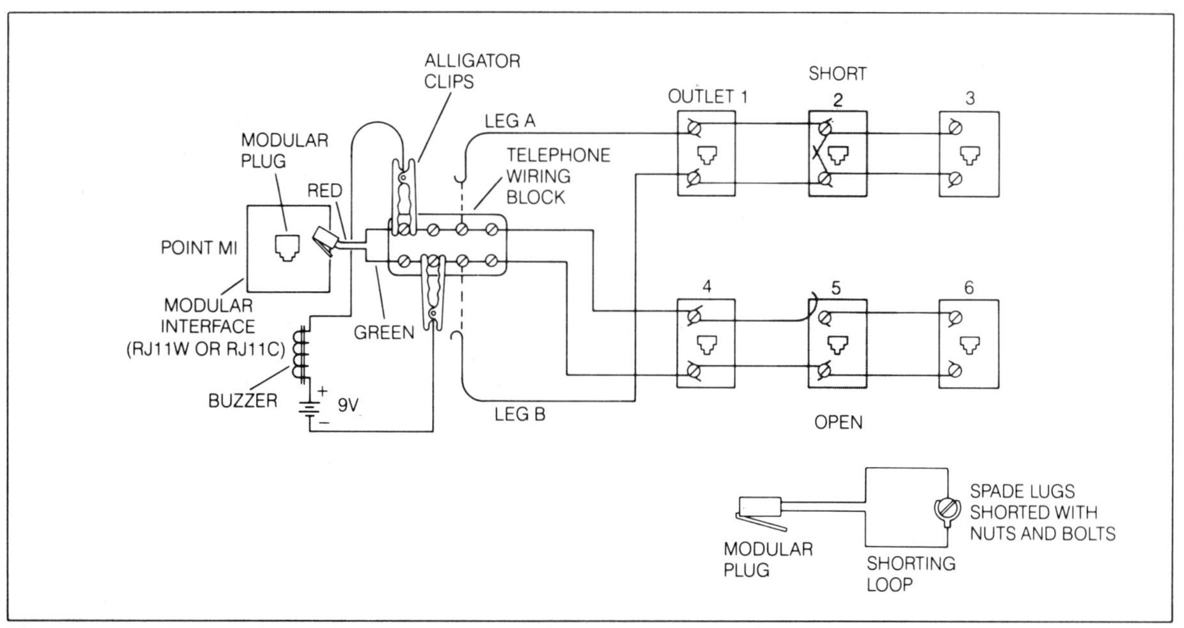

Figure 11-3.
Battery and Buzzer

Plugging in a good telephone at point MI after unplugging the new installation, has isolated the problem to the new installation. Therefore, to begin the troubleshooting, the telephone line to outlets 1 through 6 is left unplugged at point MI.

With leg A and leg B connected to the telephone wiring block, the battery and buzzer are clipped across the red and green terminals. The buzzer sounds because there is continuity and current flows in the telephone line due to the short at outlet 2. If all outlets were clear and there were no telephones in the outlets, the buzzer would not sound because the new installation telephone line is just a pair of open wires.

Isolating Leg B

Leg A is disconnected to isolate the problem and when it is disconnected the buzzer stops sounding. This indicates that the short that was on the line is in Leg A. Leg B appears to be correct, but is it?

As shown in *Figure 11-3*, a shorting loop is made from a 12″ modular-to-spade-lug line cord with the spade lugs of the red and green wires shorted by a nut and bolt. The shorting loop is plugged in outlet 6 with the battery and buzzer connected as shown in *Figure 11-3*. The buzzer does not sound. The shorting loop is plugged in outlet 5. The buzzer does not sound. In both cases the buzzer should sound because the open line is shorted by the shorting loop to produce continuity. Therefore, there is some problem with outlet 5 and 6.

This is verified by plugging the shorting loop in outlet 4. Now the buzzer sounds because the shorting loop completed the circuit so current could flow from the battery, through the buzzer, and back to the battery. The circuit open at outlet 5 prevents the current from flowing when the shorting loop is in outlet 5 and 6.

Examining the wiring on outlet 5 detects the circuit open, which when repaired, sounds the buzzer as the shorting loop is plugged in outlet 5 and 6. The open in leg B has been detected and repaired.

Isolating Leg A

Reconnecting leg A makes the buzzer sound again. Outlet 3 is examined first, no problems are detected. Outlet 2 is examined next and the short is discovered. It is repaired and the buzzer stops sounding. The system is now clear of problems, the battery and buzzer are disconnected and the modular plug connected at point MI. Plugging in the good telephone in all outlets shows that the system is operating properly.

Multi-Meter

For the person who is more technically minded, a multi-meter that measures voltage, current and resistance can be used as another way of isolating the wiring problems.

An example is shown in *Figure 11-4*. As before, by plugging a good telephone in each outlet, it was found that no outlets are working. The modular plug at point MI was removed and a telephone plugged in to verify that the line from the local telephone company is working properly. However, since the meter can measure voltage, it is just as easy to start the isolation process by measuring the voltage on the line with the telephone wiring block plugged in at point MI. The minus (−) lead of the meter is connected to the red (ring) terminal and the plus (+) meter lead is connected to the green (tip) terminal. However, in order to eliminate leg A and leg B at this time, they are disconnected from the wiring block as shown in *Figure 11-4*.

The voltage measured should be about 48 volts dc, with the red terminal negative compared to the green terminal.

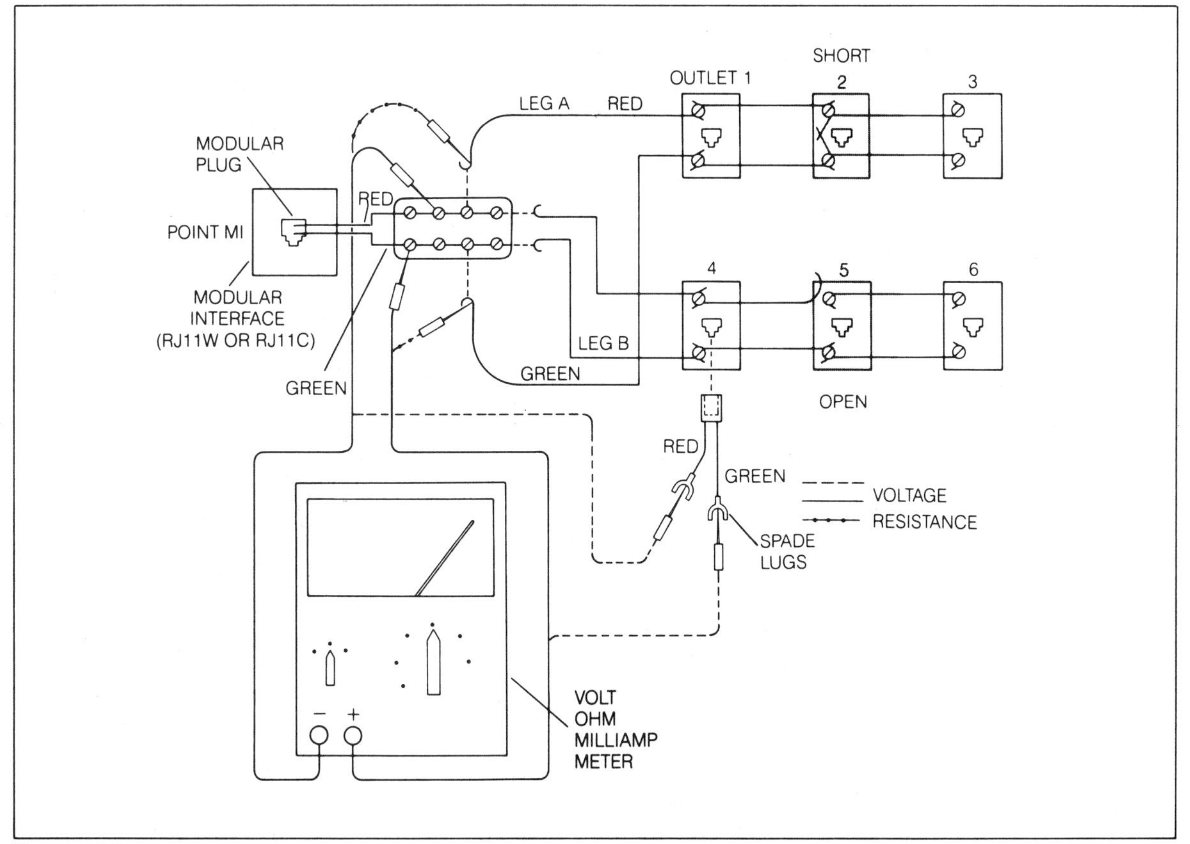

Figure 11-4.
Multi-Meter Measurements

Isolating Leg B

Now leg A is reconnected. This causes the voltage to reduce to zero because of the short in leg A. Leg A is disconnected again. However, reconnecting leg B causes no variation in voltage, it is the same as it read with leg B disconnected.

To check out leg B, the modular-to-spade-lug line cord used in *Figure 11-3* is now used as an open line rather than a shorted loop. It is plugged in outlet 4 as shown in *Figure 11-4* and the voltage measured at the red and green spade lugs. The voltage is the same as read at the wiring block so the line is good to outlet 4.

However, checking outlet 5 and outlet 6 in the same way shows that the voltage is zero at outlet 5 and outlet 6. Therefore, this isolates the circuit open in the line to outlet 5. Examining outlet 5 reveals the open line. Repairing it and plugging the modular-to-spade lug line cord in outlet 5 and outlet 6, the measured voltage is now the same as the voltage at the telephone wiring block. Leg B is repaired and working properly.

Isolating Leg A

The function of the meter is now changed to read resistance (ohms). Essentially it becomes a continuity meter like the battery and buzzer, except the indicator is the meter needle rather than the buzzer sound. To check the disconnected leg A, the meter leads are connected to the red and green open wires for leg A as shown in *Figure 11-4*. The meter indicates continuity because of the short in outlet 2. In other words , the needle deflects to indicate current flow because there is a complete circuit. The meter has a battery to supply the current.

To isolate the short, the outlet wiring is examined starting with outlet 3. When outlet 2 is reached, the short is detected and repaired and the meter no longer indicates continuity, but an open circuit. Thus, the short at outlet 2 has been cleared and the system is all ok. Leg A is reconnected and all outlets are available to operate telephones properly.

Using Extra Cable Pairs

If at any time an open is detected in a cable, remember there is at least an extra pair of wires in most cables (except for 4-conductor cables for telephones with lights) that can be used to clear the open. Substitute the yellow and black wires for the red and green wires in 4-conductor cables or another pair in other cables.

OTHER COMMON PROBLEMS

Modular Connector Rails

Modular outlets that are mounted incorrectly may have the jack positioned so the rails are on the bottom. Moisture and dirt have a chance to get between the jack rails and the plug connections to cause intermittent or no connection at all. Make sure the modular jacks are positioned so the rails are at the top.

Intermittent Contacts in Telephones

Contacts that are not making good electrical contact either cause misoperation or noise and static in the telephone.

Clicks in the receiver when dialing, incorrect numbers being dialed, and dial tones not being broken are some problems that indicate dirty contacts in the telephone.

Examples are the contacts inside the telephone set (shown in *Figure 11-5*) for pulse dialing that actually interrupt the current (S3), as well as the ones that short across the receiver (S3A) to keep out the voltage spikes caused by pulse dialing. In addition, there are the switchhook contacts (Sl and S2) that connect the transmitter and receiver and the pulse-dialing contacts to the telephone line.

Cleaning the contacts with fine sandpaper (not emery cloth) should eliminate most of the problems. A fingernail sandpaper board is a good tool for this.

Loose terminal screws on outlet connections are the main cause of loose connections. Tighten all connections securely.

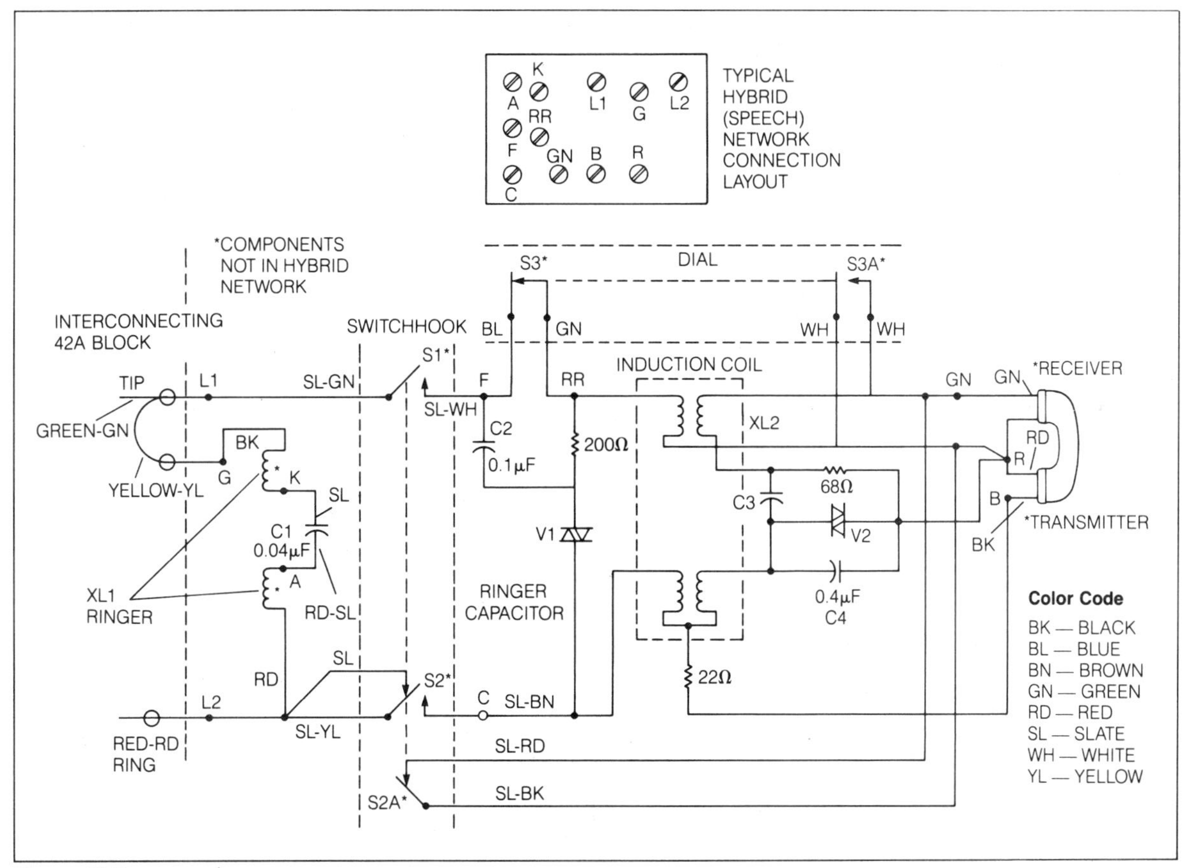

Figure 11-5.
Telephone Set Circuit Diagram

Transmitter

When a transmitter goes bad, it usually causes noise in the transmission or causes your voice to be very weak over the line. One possible cure for a transmitter is to tap the transmitter end of the handset firmly on a counter or table a couple of times. This disturbs the carbon granules in the transmitter that may have stuck together over a period of time.

Receiver

If a receiver goes bad it usually does so over a period of time. The most common indication is a distortion in the sound. An easy check is to substitute a receiver from a good telephone.

Ringer Problems

When the telephone doesn't ring, one common problem is that the ringer has been turned down by the thumbwheel or lever that adjusts the loudness of the ring. Check for a correct adjustment.

Rotary Dials

Rotary dials often go bad. If you consistently get wrong numbers as you pulse dial with a rotary dial, the problem most likely is the dial.

The dial can be removed as follows to examine and clean the contacts: Rotate the dial clockwise as far as it will go. With the dial held in this position, straighten a paper clip and insert the end in a small hole in the center ring of the dial between the 9 and the 0 finger holes. Pressing down with the paper clip as the dial is rotated clockwise will release a tab that holds the dial in place. When the tab releases, the dial will rotate clockwise, first a small amount and then, when the paper clip is removed, freely . When the dial is released, it will rotate back into its rest position, and can be lifted from the telephone set base.

After examining the contacts, cleaning them, and deciding that the dial will still work, the dial is replaced as follows: Position the dial on the telephone set base with the finger hole for the 1 digit one position further clockwise than it normally is in the rest position. In other words, the finger hole for the 2 position is now at the 1 position, the 3 position at the 2 position, etc. The finger hole for the 1 position does not line up with any digit. Press the dial in place gently, tipping it from side-to-side until it clips in place. Rotate the dial counterclockwise until it locks in place.

REFERENCE TABLE FOR TROUBLESHOOTING

The common telephone problems that arise are cataloged in *Table 11-1*. Most of the ones listed have been discussed, but there are a number that are internal to the telephone set that can affect the system operation in a number of ways. *Figure 11-5* is included and referenced where possible to provide those who are interested more technical insight into the problems. Most of the listings are problems that occur after the telephone system has been working properly.

Table 11-1.
Troubleshooting

Problem	Cords* H	Cords* L	Cause	Cure	Fig. 11-5
			Cords and their connections are common causes of problems. An X in a column means the cord indicated could cause the problem. H — Handset L — Line	Cord problems in most cases are cured by replacing the cord. Sometimes, modular plugs or jacks can be repaired.	
Dialing No Dial Tone	X	X	Cord open or not plugged in	Replace cord.	
	X	X	Modular connections defective	Replace or repair. Check wall outlets.	
			Defective switchhook (Contacts don't close)	Clear sticking buttons on cradle or button or bar that is jammed by case.	S1, S2
			Defective switchhook (Contacts don't close)	Spring broken inside telephone. Replace spring.	S1, S2
			Defective switchhook	Contacts corroded. Sandpaper.	
			Receiver defective	Substitute receiver from another telephone. If defective, replace.	
			Defective switchhook. (Contacts won't open. Receiver shorted.)	Adjust contacts.	S2A
Dial Tone Stays On			Defective rotary dial (Contacts won't open.)	Remove dial. Adjust contacts.	S3
			For tone dialing	Contact local telephone company for tone-dialing service. Reverse Green (Tip) and Red (Ring) wires on telephone set.	
			For electronic pulse dialing	Pulses per second switch was on 20 pps, should have been on 10 pps. A button contact may be defective. Try other buttons (other exchanges).	
			Ringer capacitor shorted	Replace	C1
			Filter capacitor shorted	Replace	C2
No DTMF Tones			Defective DTMF push buttons	Check push button contacts. May only be one button. To isolate, dial different numbers.	
			Defective electronics	Take to store where it was purchased for service.	
			Green (Tip) and Red (Ring) wire reversed	Reverse the green and red wires.	
Dialing Wrong Numbers			Defective rotary dial (Contacts corroded)	Remove dial, sand contacts.	S3
			Defective dial	Spring weak, replace dial.	
			Electronic pulse dialing	Pulses per second on 20 pps should be 10 pps.	

*Be aware that cords that connect handsets to telephone set bases have modular connectors that are smaller than those that connect the base to a modular outlet. Interchanging or using the wrong cord will cause problems.

Table 11-1.
Troubleshooting (continued)

Problem	Cords H	Cords L	Cause	Cure	Fig. 11-5
Dialing Clicks in Receiver			Rotary dial defective (Contacts not closing)	Remove dial, sand and adjust contacts. Some pulse dialing telephones normally click in the receiver.	S3A
Bell Taps on Dialing			Green (Tip) and Red (Ring) wires are reversed for rotary dial	Reverse the green and red wires. (Usually will solve problem)	
Ringing Dial Tone But No Ring			Ringer loudness adjustment turned down or off.	Adjust lever or thumbwheel for maximum loudness.	
			Ringer loudness adjustment turned down and telephone sitting on sound-absorbing material	Place telephone on hard, bare surface. Adjust loudness control.	
			Bell clapper jammed	Adjust case or clapper to relieve jamming.	
			Ringer capacitor open	Replace capicator.	C1
			Ringer coil open	Replace ringer.	XL1
			Defective switchhook (Contacts not closing)	Some telephone sets have one set of switchhook contacts in series with ringer coil, rather than in line L1 as shown in *Figure 11-5*. Sand contacts and adjust.	S1
Rings But Low Volume			Telephone sitting on sound-absorbing material	Place telephone on hard bare surface. Adjust loudness control.	
			Exceeding REN rating	REN greater than 5. Remove some telephones from line.	
			Bells are loose	Remove case, tighten bells, adjust loudness.	
			Bell clapper jammed	Adjust case or clapper to relieve jamming.	
Rings, You Can Hear, Can't Talk	X		Open cord to transmitter	Replace handset cord.	
			Defective transmitter	Substitute transmitter from another telephone. If defective, replace.	
Rings, You Can't Talk, Can't Hear	X		Defective handset cord	Replace cord.	
			Defective receiver (May be transmitter too)	Substitute receiver from another telephone. If defective, replace. If no transmission, do the same for transmitter.	
			Defective switchhook (Contacts won't open)	Remove case, sand and adjust contacts.	S2A
			Defective rotary dial (Contacts that should be open with dial at rest are closed)	Remove dial, sand and adjust contacts.	S3A
Transmission You Can Hear, Can't Talk	X		Open cord to transmitter	Replace handset cord.	

Table 11-1.
Troubleshooting (continued)

Problem	Cords H	Cords L	Cause	Cure	Fig. 11-5
You Can Hear, Can't Talk (Cont.)			Defective transmitter	Substitute transmitter from another telephone. If defective, replace.	
You Can Talk, Can't Hear	X		Defective handset cord	Replace cord.	
			Defective receiver	Substitute receiver from another telephone. Replace if defective.	
			Defective switchhook. (Contacts won't open)	Remove case, sand and adjust contacts.	S2A
			Defective rotary dial (Contacts that should be open with dial at rest are closed)	Remove dial, sand and adjust contacts.	S3A
Called Party Hears Distortion	X	X	Loose connections. Intermittent cord.	Replace or repair cord. Check the outlet connections.	
			Defective transmitter.	Substitute transmitter from another telephone. If defective, replace. (As stated previously, tapping the transmitter end of the handset on a hard surface several times can readjust carbon granules that might be sticking together in the transmitter and causing the distortion.)	
Cordless Telephones			Handset won't work (For handsets that contain special static discharge protection circuit)	Static discharge from walking across carpet may have activated protection circuit and turned off handset. Turn off handset for few seconds; then turn it on again. It should work now.	

Telephone Equipment Repair

FCC Part 68 — Connection of Terminal Equipment to the Telephone Network

Section 68.216 — Repair of Registered Terminal Equipment and Registered Protective Circuitry

Repair of registered terminal equipment and registered protective circuitry shall be accomplished only by the manufacturer or assembler thereof or by their authorized agent; however, routine repairs may be performed by a user, in accordance with the instruction manual if the applicant certifies that such routine repairs will not result in noncompliance with the rule and regulations in subpart D of this Part.

Rights of Telephone Company

The local telephone company may discontinue your service temporarily, if equipment that you attach to the line causes damage to the telephone network. You should be notified in advance if this happens. However, if advance notice cannot be given, you'll be notified as soon as possible and you will be given the opportunity to correct the problem. You also have a right to file a complaint with the FCC if this happens.

GLOSSARY

AC Adapter: A transformer type power supply that plugs into an ac power outlet and provides either a low ac or dc voltage to supply power to accessory equipment.

Acoustically Coupled: Equipment that is coupled to telephone lines or other equipment with a sound wave link.

Amplifier: An electronic device or circuit used to increase signal power or amplitude.

Announcement: The message that is placed on the telephone line when the telephone is answered by an answering machine.

Auto Dialing: Sending pulse-dialed numbers stored in memory on the telephone line with the push of one button.

Base Unit: The stationary unit of a cordless telephone.

Baud Rate: The rate at which modems transmit mark and space digital signals along telephone lines.

Bell Taps: Slight ringing of the bells of a telephone set as a result of reverse current through the ringer winding.

Bill of Materials: A list of items to be purchased to complete a telephone installation.

Bottom Plate: A single 2″ x 4″ wooden beam that is mounted to rough flooring to provide a base for 2″ x 4″ wall studs.

Cable: A bundled set of telephone wires. 2 pair (4 wires), 3 pair (6 wires) are common cables for a home. Apartments cables may have hundreds of wires.

Called Party: The person intended to receive a telephone call.

Calling Party: The person originating a telephone call.

Central Office: Local telephone exchange that provides telephone service to up to 9999 subscribers.

Circuit: An interconnection of electrical or electronic devices such as telephones so that when a voltage is applied a current flows in the closed circuit.

Circuit Diagram: A symbolic (schematic) description of the interconnection of electrical or telephone equipment.

Concealed Telephone Cable: Interconnecting telephone cable that is not exposed but runs inside of walls, attics and basements.

Current: The flow of electric charge (electrons) measured in amperes (see Ohm's Law).

Dial Tone: The signal received back from the central office when a telephone goes off hook to tell the caller the system is ready.

Drywall: A type of premolded wall board that substitutes for lathe and plaster for the interior of buildings.

DTMF: Dual-Tone Multi Frequency (see Tone Dialing).

Exposed Telephone Cable: Interconnecting telephone cable that is mounted on baseboards, door frames and in wall corners so it can be seen.

Exterior Wiring: Telephone lines that have their cable run on the exterior of a building.

Faceplate: The plate that fits onto an outlet box, mounting ring, or backplate that holds the telephone jack.

FCC Registration Number: The manufacturers registration number obtained from the Federal Communications Commission.

Fish Tape: Metal cable with hook on end that is used to string interconnecting cable.

Flash: A control signal placed on the line, usually by pressing a single key, that activates services like call-waiting.

Floor Plan: A building plan that shows location of existing telephones, outlets, and interconnecting cables.

4-Prong Jack: A female old style telephone outlet connector.

4-Prong Plug: A male old style telephone outlet connector.

42A Block: A common terminal block to provide interconnecting junctions of telephone line circuits.

Full-Duplex: The condition when transmitting and receiving signals on a telephone line are being sent in both directions at the same time.

Ground Line: The line that connects the protector to ground.

Half-Duplex: The condition when transmitting and receiving signals on a telephone line only occur in one direction at a time.

Handset: The part of a telephone that contains the transmitter and receiver.

Handset Cord: The cord that runs between the handset and the telephone set.

Initial Input Termination: The connector to which the telephone line is connected when it first enters a home, apartment or small business.

LDT: (Long Distance Tone) A control signal that switches pulse dialing to tone dialing for access to data and other telecommunication services.

Leg: A branch of a telephone circuit.

Line Cord: The cord that connects the telephone set to the telephone outlet.

Line Coupler: A duo-jack used to lengthen telephone extension lines by plugging in modular plugs.

Local Loop: Telephone circuit from a given telephone set to the central office.

"Looped" Connection: The name given to connections made when bare cable wires are wrapped around terminals while keeping the wires continuous.

Mark: The higher tone frequency of two tones that are used to identify digital signal levels on telephone lines.

Memory: Electronic circuits in telephones or auto dialers that store information.

Microphone: A unit that converts sound waves into electrical signals by magnetic coupling. Used as a telephone transmitter.

Modem: An accessory to couple digital signals from computers and terminals to telephone lines. Short for Modulation and Demodulation.

Modular Connector Rails: The small wires inside a modular jack and on the outside of a modular plug that make the electrical connection when plug and jack are put together.

Modula Interface: A modular outlet with a modular jack that initially terminates the telephone line from the protector. Used as a quick-disconnect connection point.

Modular Jack: The female connection of a modular telephone outlet.

Modular Plug: The male connector of a modular telephone connection that plugs in a modular jack.

Modular Wall Telephone Plate: The plate used to interconnect a modular wall telephone.

Moly Bolt: A trade name for a particular kind of anchor that inserts into a hole in drywall and expands metal ears behind the drywall to secure the mounting.

Multi-Meter: Electrical test instrument that measures voltage, current and resistance.

National Electric Code: A code that outlines requirements to help install safe, reliable electrical or telephone wires.

Notes of Caution: Installation precautions that should be heeded.

Off Hook: A condition where current flows in the local loop telephone circuit.

Ohm's Law: The voltage E, in volts, applied to a circuit is equal to the current I that flows, in amperes times the resistance, in ohms, of the circuit, i.e. $E = IR$.

Open: An open wire of a telephone line circuit such that current will not flow in the circuit.

Outlet: A faceplate that provides connection to a telephone line.

Outlet Box: The electrical outlet box used for mounting telephone outlets.

Paging: Alerting a person with a radio signal to call the office, home, or seleted number to receive a message.

Pause: A programmed break in a dialing sequence. For example, inserting 9 to gain access to an outside line.

Party Line: A telephone line whose local loop circuit is shared by a number of parties.

Plaster and Lathe: A type of construction of interior walls of buildings that has wooden slats called lathe for the base of plaster walls.

Plastic Screw Anchor: Inserts that are pressed into drywall to provide stable mounting of screws so the screws will be anchored rather than pull loose.

Plumbbob: The weight that is placed on the end of a string to produce a vertical line for reference to squareness.

Prewiring: Stringing telephone cables, electrical wiring, appliance control wiring while a home, apartment building or condominium is under construction.

Privacy: A feature, usually initiated by pressing a single key, that disconnects the microphone input of a feature telephone so that the caller cannot hear the called party's conversation.

Protector: A unit that protects telephone lines from high voltage due to lightning bolts.

Pulse Dialing: (also Rotary Dialing) Providing the telephone system with the telephone number called by turning current off or on in the local loop.

Pulse Train: A series of current pulses that appear on a local loop for pulse dialing.

Receiver: The part of the handset that converts electrical signals to sound. It is the part of the system that is receiving information.

REN (Ringer Equivalence Number): A standard ringing power unit. 5 REN normally is supplied by a telephone company.

Resistance: The resistance in ohms to the flow of current in amperes caused by an electrical force in volts being applied to a circuit (See Ohm's Law).

Ring: The name of one conductor of a telephone line pair, identified by R. Most commonly the red wire, and the most negative of the two.

Ringer: The mechanism that does the ringing.

Ringing: Applying a 85-90v ac signal to the telephone line to alert the called party that a call is waiting.

RS-232: The specification telling the value of digital signals levels, the particular way they are arranged, and how they must occur in time to send digital information from one piece of equipment to another.

Service Charge: The fee paid for maintenance and repair.

Short: The direct connection together of the two wires of a telephone line pair.

Shorting Loop: A test loop used in troubleshooting a telephone circuit that inserts a short into the circuit.

Sidetone: The part of the speech into the transmitter that is fed to the receiver.

Single-line Tap: Isolating a single telephone line from a multi-line system of telephone lines.

Snap-on Connectors: Terminal connectors that snap over screw terminals in order to make electrical connection.

Space: The lower tone frequency of two tones that are used to identify digital signal levels on telephone lines.

Speaker: A unit that converts electrical signals to sound waves by use of a voice coil in a magnetic field — the output unit of a sound amplifier.

Speech Network: The circuit inside the telephone that sets the amount of sidetone.

Switchhook: The spring loaded mechanism that closes switch contacts to complete the local loop circuit from the telephone set to the central office.

Telephone Accessory: Equipment other than telephone sets that is connected to or used with telephone lines and the telephone system.

Telephone Service Entrance: The location in the exterior wall of a building where the telephone line enters the building.

Telephone Wiring Block: A wiring block used for a terminal block to interconnect telephone lines that junction at a particular point. Particularly useful to provide the quick-disconnect to the modular interface, the initial termination of the telephone line from the protector.

Tip: The name of one conductor of a telephone line pair, identified by T. Most commonly the green wire, and the most positive of the two.

Tone Dialing: Providing the telephone system with the telephone number called by sending combination tones to the central office.

Top Plate: Two 2" x 4" wooden beams that run across the top of wall studs.

Transmitter: The part of the handset that converts speech to electrical signals. It is the part of the system that is sending information.

Trouble Shooting: Finding and repairing faults that occur in telephone equipment or interconnecting lines in a telephone system. Commonly called debugging.

Type A Line: Single party ringing line with a ringing frequency of 20 Hz and 30 Hz.

Voice Band: A band of frequencies from 200 Hz to 6,000 Hz that is typical for voice signals.

Voltage: A measure of the electrical force in volts that causes current to flow in a circuit (See Ohm's Law).

INDEX

Acoustically Coupled: **10**-12
Adapter
AC: **8**-9, 10; **10**-4, 6, 8, 9
Cord: **4**-7, 11; **7**-3
Answerer: **8**-6
Amplifier: **1**-5; **8**-3; **10**-14, 15
Announcement: **1**-5, **10**-9, 10
Area Codes: **6**-18
Auto Dialer: **6**-16; **8**-3; **10**-3, 4
Base Unit: **10**-5, 6
Baud Rate: **10**-12, 13
Bell Taps: **11**-13
Bill of Materials: **9**-7, 8
Bottom Plate: **5**-10, 13; **6**-9, 10
Business Installtions
Single-Line: **8**-2
2-Line: **8**-5
4-Line: **8**-7
Cable
Attic: **5**-9
Cabinet Runs: **5**-5
Closet Runs: **5**-6
Color Code: **5**-2
Concealed: **5**-4, 7, 8, 9, 10, 11; **6**-8, 9; **7**-5, 7
Distribution: **1**-6; **2**-11
Exposed: **5**-3, 4, 5; **6**-5, 7
Extension: **3**-14; **6**-3
Exterior: **5**-14
Modular: **3**-8, **5**-3
Round: **3**-9; **5**-3; **9**-12
Under Carpet: **5**-5
Central Office: **1**-6
Connector Cap: **4**-15
Dial Tone: **1**-11
DTMF: **1**-15
Faceplate: **1**-9, 18; **2**-8, 10; **4**-9, 10, 11, 15, 16
FCC
Registration No.: **3**-2
Regulations: **1**-2
Fish Tape: **5**-8, 9, 10; **6**-7, 10, 11; **7**-5
Floor Plan: **2**-11, 12; **3**-6; **4**-3; **6**-3; **7**-4; **9**-5, 6
Four Prong (4-prong)
Jack: **1**-18, **2**-8; **4**-7, 9, 11
Plug: **1**-18; **2**-8; **4**-7, 11; **7**-3
42A Block: **2**-6; **4**-2, 3, 4; **5**-9, 13, 15; **6**-5, 6; **7**-4, 5, 9; **9**-8
Full Duplex: **10**-13
Hand Set: **1**-11, 12, 13
Initial Input Termination
See 42A Block
See Modular Interface
Line Coupler: **3**-14
Line Tap: **8**-3, 4
Local Loop: **1**-6
"Looped" Connection: **1**-10; **2**-11; **9**-5, 6
Mark: **10**-12
Memory **1**-4; **10**-3
Microphone: **10**-15
Modem: **10**-10, 11, 12, 13, 14
Modular Interface: **1**-9; **2**-7; **5**-3
Modular Jack
Duplex: **6**-3, 4; **7**-3; **8**-3, 6, 9; **10**-4
Spadelugs: **2**-7; **4**-6
Snap-on: **2**-7; **4**-6
Variety: **1**-18; **2**-7, 9, 10; **3**-7, 8, 9, 13; **4**-6, 7, 10; **5**-4, 5, 6, 7, 8, 12
Modular Plug: **1**-18; **2**-9, 10; **3**-7,9, 13, 14; **4**-6, 7
Modular Wall Plate: **4**-15, 16, 17; **5**-6
Multi-Line Controller: **8**-5
Multi-Meter:**11**-8
National Electric Code: **9**-2
Notes of Caution: **3**-4
Open: **11**-3, 5, 6, 8
Outdoor Ringer: **9**-8, 11; **10**-18
Outlet — Modular
Over 42A: **2**-10; **4**-6
Rectangular: **1**-9, 18; **2**-10, 3-7; **5**-4,6, 12; **6**-4; **8**-3; **9**-16
Round: **2**-10; **5**-5, 6, 7, 8; **7**-5, 8, 9, **9**-9
Surface-Mounted: **2**-10; **3**-8, 9; **4**-6, 7,11; **5**-4; **6**-3, 4
Outlet — Old Style
Rectangular: **1**-18; **2**-8; **4**-9, 11; **7**-3
Round: **2**-8; **7**-4
Outlet Box: **4**-9, 10; **5**-9, 10, 12, 13; **6**-7, 11; **9**-5
Pocket Tone Dialer: **10**-16, 17
Prewiring: **9**-12, 13, 14, 15, 16, 17, 18
Protector: **1**-6; **2**-2, 3, 4, 5
Pulse Dialing: **1**-14; **3**-3; **10**-3
Receiver: **1**-13
Recording Controller: **8**-10
Remote Control-Answerer: **10**-10
REN: **3**-2, 3
Ring: **1**-10; **5**-2; **6**-14; **11**-10
Ringer: **1**-16, 17
Ringer Silencer: **10**-17, 18
RJ11: **6**-15
RJ14: **6**-15
RS-232: **10**-13
Service Charge: **1**-3
Short: **11**-3, 4, 5, 6, 8
Shorting Loop: **11**-6
Sidetone: **1**-16
Single-Line Tap: **8**-4
Single-Line to 2-Line: **6**-17
Snap-on Connectors: **2**-7; **4**-6
Space: **10**-12, 13
Speaker: **10**-9
Speech Network: **1**-16; **4**-14
Switchhook: **1**-11; **11**-10, 12, 13, 14
Telephone
Amplifier: **1**-1; **8**-3, **10**-14, 15
Answerer: **1**-1; **8**-3, 6, 7, 9; **10**-7, 8, 9, 10
Conference Amplifier: **1**-5; 8-3
Contemporary Wall: **3**-11, 14
Cordless: **1**-1; **10**-5, 6, 7
Desk (Table): **1**-11; **3**-8, 9; **4**-11; **8**-3; **10**-4, 9
Extension: **1**-11; **3**-7, 4-3, 4
Flasher: **9**-11; **10**-15, 16
4-Line: **8**-7, 8, 9
Lighted: **4**-4
Service Entrance: **9**-14
Specialty: **1**-4
Speakerphone: **6**-12, 16; **8**-6, 9; **10**-8
Type A Line: **8**-2
2-Line: **6**-13, 17; **8**-5, 6, 8
Wall: **1**-11; **2**-9; **3**-10-13; **4**-13,14
Wiring Block: **5**-3, 13; **9**-8, 9; **11**-3, 5, 6, 8
Tip: **1**-10; **5**-2; **6**-14; **11**-10
Tone Dialing: **1**-14, 15; **2**-12; **3**-3, 12; **10**-5
Tools: **3**-5, 6
Top Plate: **5**-9, 10, 13; **9**-16
Transmitter: **1**-12
Voice Band: **10**-10

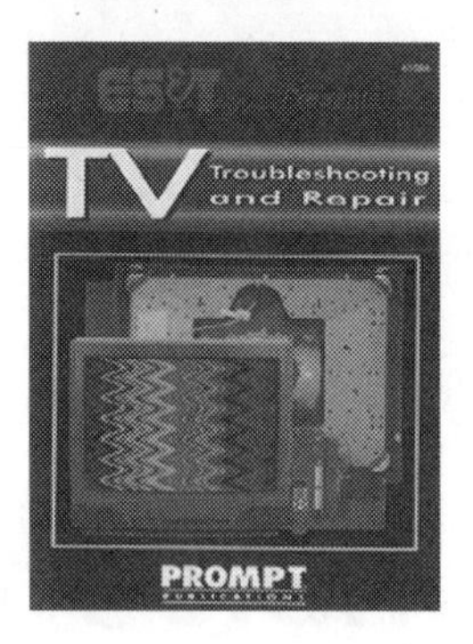

ES&T Presents TV Troubleshooting & Repair

Electronic Servicing & Technology Magazine

TV set servicing has never been easy. The service manager, service technician, and electronics hobbyist need timely, insightful information in order to locate the correct service literature, make a quick diagnosis, obtain the correct replacement components, complete the repair, and get the TV back to the owner.

ES&T Presents TV Troubleshooting & Repair presents information that will make it possible for technicians and electronics hobbyists to service TVs faster, more efficiently, and more economically, thus making it more likely that customers will choose not to discard their faulty products, but to have them restored to service by a trained, competent professional.

Originally published in *Electronic Servicing & Technology*, the chapters in this book are articles written by professional technicians, most of whom service TV sets every day.

Video Technology
226 pages • Paperback • 6 x 9"
ISBN: 0-7906-1086-8 • Sams: 61086
$18.95 ($25.95 Canada) • August 1996

ES&T Presents Computer Troubleshooting & Repair

Electronic Servicing & Technology

ES&T is the nation's most popular magazine for professionals who service consumer electronics equipment. PROMPT® Publications, a rising star in the technical publishing business, is combining its publishing expertise with the experience and knowledge of *ES&T's* best writers to produce a new line of troubleshooting and repair books for the electronics market. Compiled from articles and prefaced by the editor in chief, Nils Conrad Persson, these books provide valuable, hands-on information for anyone interested in electronics and product repair.

Computer Troubleshooting & Repair is the second book in the series and features information on repairing Macintosh computers, a CD-ROM primer, and a color monitor. Also included are hard drive troubleshooting and repair tips, computer diagnostic software, networking basics, preventative maintenance for computers, upgrading, and much more.

Computer Technology
288 pages • Paperback • 6 x 9"
ISBN: 0-7906-1087-6 • Sams: 61087
$18.95 ($26.50 Canada) • February 1997

CALL 1-800-428-7267 TODAY FOR THE NAME OF YOUR NEAREST PROMPT PUBLICATIONS DISTRIBUTOR

Theory & Design of Loudspeaker Enclosures

Dr. J. Ernest Benson

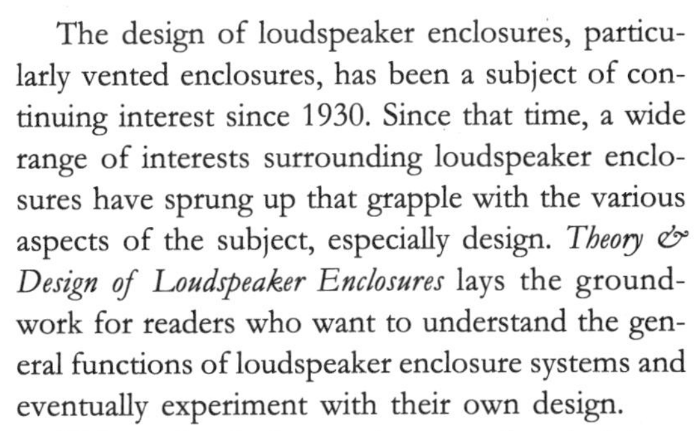

The design of loudspeaker enclosures, particularly vented enclosures, has been a subject of continuing interest since 1930. Since that time, a wide range of interests surrounding loudspeaker enclosures have sprung up that grapple with the various aspects of the subject, especially design. *Theory & Design of Loudspeaker Enclosures* lays the groundwork for readers who want to understand the general functions of loudspeaker enclosure systems and eventually experiment with their own design.

Written for design engineers and technicians, students and intermediate-to-advanced level acoustics enthusiasts, this book presents a general theory of loudspeaker enclosure systems. Full of illustrated and numerical examples, this book examines diverse developments in enclosure design, and studies the various types of enclosures as well as varying parameter values and performance optimization.

Audio Technology
244 pages • Paperback • 6 x 9"
ISBN: 0-7906-1093-0 • Sams: 61093
$19.95 ($26.99 Canada) • August 1996

Making Sense of Sound

Alvis J. Evans

This book deals with the subject of sound — how it is detected and processed using electronics in equipment that spans the full spectrum of consumer electronics. It concentrates on explaining basic concepts and fundamentals to provide easy-to-understand information, yet it contains enough detail to be of high interest to the serious practitioner. Discussion begins with how sound propagates and common sound characteristics, before moving on to the more advanced concepts of amplification and distortion. *Making Sense of Sound* was designed to cover a broad scope, yet in enough detail to be a useful reference for readers at every level.

Alvis Evans is the author of many books on the subject of electricity and electronics for beginning hobbyists and advanced technicians. He teaches seminars and workshops worldwide to members of the trade, as well as being an Associate Professor of Electronics at Tarrant County Junior College.

Audio Technology
112 pages • Paperback • 6 x 9"
ISBN: 0-7906-1026-4 • Sams: 61026
$10.95 ($14.95 Canada) • November 1992

CALL 1-800-428-7267 TODAY FOR THE NAME OF YOUR NEAREST PROMPT PUBLICATIONS DISTRIBUTOR

The Video Book

Gordon McComb

Televisions and video cassette recorders have become part of everyday life, but few people know how to get the most out of these home entertainment devices. *The Video Book* offers easy-to-read text and clearly illustrated examples to guide readers through the use, installation, connection, and care of video system components. Simple enough for the new buyer, yet detailed enough to assure proper connection of the units after purchase, this book is a necessary addition to the library of every modern video consumer. Topics included in the coverage are the operating basics of TVs, VCRs, satellite systems, and video cameras; maintenance and troubleshooting; and connectors, cables, and system interconnections.

Gordon McComb has written over 35 books and 1,000 magazine articles, which have appeared in such publications as *Popular Science*, *Video*, *PC World*, and *Omni*, as well as many other top consumer and trade publications.

Video Technology
192 pages • Paperback • 6 x 9"
ISBN: 0-7906-1030-2 • Sams: 61030
$16.95 ($22.99 Canada) • October 1992

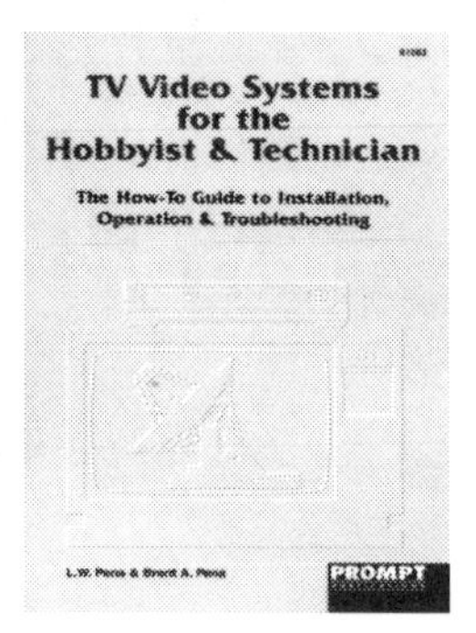

TV Video Systems

L.W. Pena & Brent A. Pena

Knowing which video programming source to choose, and knowing what to do with it once you have it, can seem overwhelming. Covering standard hard-wired cable, large-dish satellite systems, and DSS, *TV Video Systems* explains the different systems, how they are installed, their advantages and disadvantages, and how to troubleshoot problems. This book presents easy-to-understand information and illustrations covering installation instructions, home options, apartment options, detecting and repairing problems, and more. The in-depth chapters guide you through your TV video project to a successful conclusion.

L.W. Pena is an independent certified cable TV technician with 14 years of experience who has installed thousands of TV video systems in homes and businesses. Brent Pena has eight years of experience in computer science and telecommunications, with additional experience as a cable installer.

Video Technology
124 pages • Paperback • 6 x 9"
ISBN: 0-7906-1082-5 • Sams: 61082
$14.95 ($20.95 Canada) • June 1996

CALL 1-800-428-7267 TODAY FOR THE NAME OF YOUR NEAREST PROMPT PUBLICATIONS DISTRIBUTOR

PROMPT
PUBLICATIONS

The Microcontroller Beginner's Handbook

Lawrence A. Duarte

Microcontrollers are found everywhere — microwaves, coffee makers, telephones, cars, toys, TVs, washers and dryers. This book will bring information to the reader on how to understand, repair, or design a device incorporating a microcontroller. *The Microcontroller Beginner's Handbook* examines many important elements of microcontroller use, including such industrial considerations as price vs. performance and firmware. A wide variety of third-party development tools is also covered, both hardware and software, with emphasis placed on new project design. This book not only teaches readers with a basic knowledge of electronics how to design microcontroller projects, it greatly enhances the reader's ability to repair such devices. Lawrence A. Duarte is an electrical engineer for Display Devices, Inc. In this capacity, and as a consultant for other companies in the Denver area, he designs microcontroller applications.

Electronic Theory

240 pages • Paperback • 7-3/8 x 9-1/4"
ISBN: 0-7906-1083-3 • Sams: 61083
$18.95 ($25.95 Canada) • July 1996

Schematic Diagrams

J. Richard Johnson

Step by step, *Schematic Diagrams* shows the reader how to recognize schematic symbols and determine their uses and functions in diagrams. Readers will also learn how to design, maintain, and repair electronic equipment as this book takes them logically through the fundamentals of schematic diagrams. Subjects covered include component symbols and diagram formation, functional sequence and block diagrams, power supplies, audio system diagrams, interpreting television receiver diagrams, and computer diagrams. *Schematic Diagrams* is an invaluable instructional tool for students and hobbyists, and an excellent guide for technicians.

J. Richard Johnson has written numerous books covering all aspects of electronics. He worked for many years as an engineer for Bell Laboratories, and was also the managing editor of *Radio Maintenance* magazine. He currently works as a freelance writer within the electronics field.

Electronic Theory

196 pages • Paperback • 6 x 9"
ISBN: 0-7906-1059-0 • Sams: 61059
$16.95 ($22.99 Canada) • October 1994

CALL 1-800-428-7267 TODAY FOR THE NAME OF YOUR NEAREST PROMPT PUBLICATIONS DISTRIBUTOR